GM G-Body
PERFORMANCE UPGRADES
1978-1987

Joe Hinds

CarTech®

CarTech®

838 Lake Street South
Forest Lake, MN 55025
Phone: 651-277-1200 or 800-551-4754
Fax: 651-277-1203
www.cartechbooks.com

© 2013 by Joe Hinds

All rights reserved. No part of this publication may be reproduced or utilized in any form or by any means, electronic or mechanical, including photocopying, recording, or by any information storage and retrieval system, without prior permission from the Publisher. All text, photographs, and artwork are the property of the Author unless otherwise noted or credited.

The information in this work is true and complete to the best of our knowledge. However, all information is presented without any guarantee on the part of the Author or Publisher, who also disclaim any liability incurred in connection with the use of the information and any implied warranties of merchantability or fitness for a particular purpose. Readers are responsible for taking suitable and appropriate safety measures when performing any of the operations or activities described in this work.

All trademarks, trade names, model names and numbers, and other product designations referred to herein are the property of their respective owners and are used solely for identification purposes. This work is a publication of CarTech, Inc., and has not been licensed, approved, sponsored, or endorsed by any other person or entity. The publisher is not associated with any product, service, or vendor mentioned in this book, and does not endorse the products or services of any vendor mentioned in this book.

Edit by Paul Johnson
Layout by Monica Seiberlich

ISBN 978-1-61325-494-3
Item No. SA246P

Library of Congress Cataloging-in-Publication Data

Hinds, Joe.
  GM G-body performance upgrades 1978-1987 / by Joe Hinds.
    pages cm
  ISBN 978-1-61325-032-7
  1. General Motors automobiles–Maintenance and repair. 2. General Motors automobiles–Performance. I. Title.

TL215.G4H56 2013
628.28'722–dc23

2013002375

Written, edited, and designed in the U.S.A.
Printed in the U.S.A.
10 9 8 7 6 5 4 3

Front Cover: Scott Walkowiak built this 1986 Buick Regal T-Type into a genuine Pro Touring car. The car, known as "GNXray," carries Detroit Speed front and rear suspension components. In fact, this car was the test best for DSE when it developed its G-Body suspension components. The 231-ci LC2 turbo engine is equipped with a Precision TE44 turbo that supplies up to 24 psi of boost.

Title Page: Dan Howe's 1984 Monte Carlo was tearing up the track at Optima Batteries OUSCI 2012. This G-Body has a complete Schwartz Performance chassis and is powered by a 500-hp LS1 backed by a 4L60E automatic transmission. (Photo Courtesy Tony Huntimer)

Back Cover Photos

Top Left: During the project car build-up of the GNXcess, we mocked up all the components to verify the fit. While it's not essential, it's helpful to see how the parts will all come together and see how it would look assembled. Upper and lower STRONGarms, Belltech Blazer 2WD spindles and hubs, and coil-over shocks have been installed. In addition, a 525-hp LS3 crate engine from Chevrolet Performance has been installed.

Top Right: The B-Body spindle and caliper were installed using the "old school" method for installing larger discs on a G-body. It does give a taller spindle, which is desirable, but it creates a really bad angle on the tie rods and requires a longer upper A-arm. This one is installed on an early A-Body.

Middle Left: This is a 12-bolt Chevy axle assembly with a Detroit Speed & Engineering cast-aluminum cover. For easy identification at a glance, look for two horizontal bolts at the top and bottom of the housing.

Middle Right: Baer brakes front and rear also and the larger mono-block calipers are fitted to the T-Type GNXray. The mono-block calipers are more rigid than a typical cast caliper because they are machined out of a solid block of aluminum. The wheels are from CCW Classics.

Bottom Left: Hydra-Tech offers this aftermarket version of the Hydro-boost, using a late-model Corvette master cylinder and a Baer proportioning valve. A complete system retails for about $1,000.

Bottom Right: Ben Meissner built the very simple braces that attach the lower control arm mounting points to the uppers and the crossmember. The crossmember contains the driveshaft and strengthens the lower control arm mounts. All of these are bolt-in parts, so that they could be easily removed for service if needed. (Photo Courtesy Ben Meissiner)

CONTENTS

Acknowledgments 4

Chapter 1: Evolution not Revolution: G-Body History 5
    A-Body Becomes G-Body 5
    Chevrolet 6
    Pontiac 10
    Oldsmobile 10
    Buick 11

Chapter 2: On the Hunt: Finding Your Ideal G-Body 15
    Choosing a Model 15
    Setting Performance Goals 16
    Creating a Budget 17
    General Inspection 20
    Rust Issues 25
    Body Damage 28
    Chassis Considerations 28

Chapter 3: Maximizing Stopping Power:
    Stock and Aftermarket Brakes 29
    Hydroboost Brake System 30
    Powermaster Brake System 31
    Brake Upgrades 32
    Spindles 33
    Brake Proportioning Valve 36
    Brake Lines 37
    Brake Conversions 42
    Aftermarket Brake Systems 43

Chapter 4: Getting a Grip: Front Suspension Performance .. 47
    Component Options 47
    Stock Control Arms 48
    Aftermarket Control Arms 51
    Sway Bars 54
    Shocks 56
    Springs 57

Chapter 5: Handling on Rails: Steering System Upgrades 60
    Steering Box 60
    Rack-and-Pinion Conversion 62
    Power Steering Hose 67
    Power Steering Pump 67

Chapter 6: Hooking Up:
    High-Performance Rear Suspension 70
    Control Arms 70
    Bushings 71
    Sway Bars 73
    Shocks 73
    Air Bag Suspension 73

Chapter 7: Building a Strong Foundation:
    Chassis and Frame Upgrades 77
    Separate Frame from Body 77
    Frame Rail Boxing 79
    Roll Bar and Roll Cage Installation 79
    Crossmembers 83
    Wheel Clearance 85
    Aftermarket Chassis 88

Chapter 8: Putting the Power Down:
    High-Performance Differentials and Axles 90
    GM 10- or 12-Bolt 91
    Ford 9-Inch 92
    Dana 60 93
    Budget Option 93
    Stock Axle Assembly Upgrades 94
        Ring-and-Pinion Gears 95
        Pinion Supports and Yokes 95
        Axle Shafts 95
        Wheel Studs 96
        Limited-Slips 96
        Detroit Lockers 96
        Spools 97
        Differential Covers 97
        Back Braces 98
        Gear Cases 98

Chapter 9: Get Into Gear: Manual
    and Automatic Transmissions 100
    Powerglide 100
    Turbo Hydra-Matic 100
    New-Generation Transmissions 103
    OEM Manual Transmissions 104
    Aftermarket Manual Transmissions 105

Chapter 10: A High-Performance Trans-Formation:
    Manual Transmission Swaps 107
    Transmission Fit 109
    Clutch Pedal 109
    Driveshaft 111
    Speedometer Drive 111
    Shifter 111

Chapter 11: A Higher Power: Engine Swaps 112
    Traditional V-8 Power 113
    LS Series 116
    Engine Mounts 119
    Wiring and Engine Management 120
    Exhaust 121
    Intake 121
    Carburetor or EFI 123
    Fuel System 125
    Cooling System 129
    Swap Kits 130
    Gauges 132
    Accessories 133

Chapter 12: Driving in the Lap of Luxury:
    Performance Interior Upgrades 135
    Gauges 135
    Wiring Harness 138
    Steering Wheel 138
    Shifter 139
    Seats 140
    Safety Restraints 142

Source Guide 144

DEDICATION

To my wife, Nancy. Without her love and support, my shop, Bulldawg Musclecars, and this book, would not exist.

ACKNOWLEDGMENTS

Since the day my dad and I drove home my first car, a 1957 Chevy Bel Air, back in the spring of 1985, I have been obsessed with cars. For most of my life it has been a hobby, and for the last three years it has been my full-time business, Bulldawg Musclecars. I have learned so much during those 28 years; not only about the cars themselves, but about myself, and what is really possible with hard work and determination.

First, I'd like to thank my late father, since he helped me to purchase that 1957 Chevy that started it all, and supported me in my hobby (and later, my business) until his passing in May 2012. I miss him every day.

Many others helped along the way as I delved into the world of hot rodding: Jack Doss, a great friend to this day, who took the time to help a teenager learn how to work on his own car, and gave me the confidence to take on any project; Mike Buice, who always helped me to find that odd part I needed; Year One Inc., where I worked and learned an appreciation for cars other than Chevrolets, and made a lot of great friends; Jimmy Killian, who helped take my performance obsession to the next level; David Sloan, of Road Killer Kustoms, for his friendship and inspiration to make this hobby my career; Chris Johnson, Curt Johnson (no relation), and Jeremy Hale, whose loyalty, friendship, and hard work made the project 1983 Malibu wagon, *Grocery Getter*, a reality; and the late Tom Gerrard, who owned and provided the vision for the *Grocery Getter*. It was a pleasure getting to know Tom, even though it was for just a short while, and to see his smile on the Hot Rod Magazine Power Tour of 2011.

I have documented my work on several online forums (pro-touring.com, gbodyforum.com, and malibu racing.com) but this is my first serious attempt at automotive writing. I'd like to thank all the people who helped make this work possible: Phil Brewer and Jeff Georges of BRP/Musclerods; Scott Walkowiak of GNS Peformance; Brett Voelkel of RideTech; Rick and Kristi Bejarano of METCO Motorsports; Jeff and Dale Schwartz of Schwartz Performance; Chris Alston ChassisWorks; Ben Meissner of Street Rod Design and Bulldawg Musclecars; Philip Wigington of Hedman Hedders; Jeff Tate of Day's Chevrolet; Doug Lutes; Dan and Josh Howe; CarTech editor Scott Parkhurst, for having the confidence in my G-Body knowledge and writing skills to offer me this opportunity, and getting me started; CarTech editor Paul Johnson; and everyone at CarTech Books.

CHAPTER 1

EVOLUTION NOT REVOLUTION
G-BODY HISTORY

To understand how the G-Body came about, you need to look at its predecessor, the A-Body. In the form of the Pontiac GTO, Olds 4-4-2, Buick Grand Sport, and Chevelle SS, the A-Body formed GM's front line in the muscle car wars of the 1960s. These cars and their less performance oriented versions were some of the most popular, best handling, and best performing cars that General Motors had built up to that time. They built a following of loyal enthusiasts that is strong to this day.

In 1973, the chassis received a much-needed redesign, and it benefitted from the excellent suspension geometry of the second-generation F-Body (Camaro/Firebird), introduced in 1970. The cars themselves, however, never garnered the attention from enthusiasts as their predecessors did. A combination of increased weight, hideous government-mandated 5-mph bumpers, controversial styling, and lackluster powertrains have resulted in their being ignored by the mainstream as a choice for a project car, though they are increasing in popularity.

A-Body Becomes G-Body

In 1978, the A-Body was downsized for increased mileage and performance. Styling improved as well, and it was a significant step ahead of the land barges of 1973–1977. General Motors used lighter-gauge stampings, plastic, and aluminum to substantially reduce weight. In 1981, the A-Body designation was taken from the midsize rear-wheel-drive models, and reassigned to a new, front-wheel-drive chassis. From that point forward, the A-Body was called the G-Body.

Although the power choices (mostly V-6s and small V-8s) were uninspiring, the A/G-Body (I use the term G-Body, for simplicity's sake) was an immediate hit with the performance market, even though it was much later in the car's timeline when it received respectable power from

This Monte was originally blue, but the current hue is much darker and deeper than the original paint. Note the Baer brakes visible between the wheel spokes. (Photo Courtesy Rick and Kristi Bejarano)

CHAPTER 1

Early G-Bodies, including the El Camino, had vinyl rather than plastic or cloth door panels. The vinyl and plastic interior parts can easily be refinished for a new appearance at very little cost.

This El Camino owned by Nick Freiser doesn't have the typical SS-style front fascia. Instead, the factory fascia and Lauren Engineering bumpers were painted in a monochromatic scheme that is as popular today as it was in the 1980s and 1990s. The El Camino, like other G-Body cars, is an ideal platform for a variety of powertrain and suspension chassis upgrades.

the factory. The platform was similar enough to previous offerings in regard to the suspension, and unlike with today's cars, an engine swap was an easy weekend project.

The G-Body (and the A-Body before it) was offered as a two-door coupe or sedan, four-door sedan, four-door wagon, and a utility version (the Chevrolet El Camino and GMC Caballero). Although each manufacturer had its own styling and little interchange among external body panels and trim, under the skin they are all very similar. Because of this, the suspension, brake, drivetrain, and even interior items can often be swapped from one division's car to another. This makes the job of rebuilding one of these cars relatively simple, especially if you aren't concerned with originality.

In addition to the interchangeability within the G-Body line, there is plenty of interchangeability with other GM cars, made both prior to and since the G-Body. You can put a 455 Oldsmobile in your 307-powered Regal, and mostly do it with the parts you already have, including the transmission. Or you can upgrade to something completely modern, such as an LS7 from a late-model Corvette. The boundaries for these cars are nearly limitless, hampered only by imagination and budget.

Chevrolet

Chevrolet offered two distinct models on the G-Body chassis, the Malibu and the Monte Carlo. The Malibu was the more utilitarian of the two. Some early models were even equipped with the Saginaw 3- or 4-speed manual transmissions (mechanical clutch linkage), rubber floor mats, and roll-up windows. Since its release, the Malibu has been a drag racer's favorite, and it is common at drag strips across the country. The El Camino and GMC Caballero (the only GMC G-Body; an El Camino clone) were based on the Malibu, as were the Malibu wagon and all other G-Body wagons. The front sheet metal and interior trim

This nice Malibu has a very sinister look, and lots of power thanks to the LS engine swap.

GM G-BODY PERFORMANCE UPGRADES 1978–1987

on every other division's wagon were different, but all were basically Malibus from the firewall back.

The Monte Carlo was a slightly larger, more upscale model than the Malibu, and it had a much broader appeal. It was one of the most popular cars in Chevrolet history. While not embraced by the street machine crowd at first, it was always popular with the low-rider crowd. In 1983, the Monte Carlo became the basis for the only Chevrolet G-Body marketed as a performance car, the Monte Carlo SS. More of an appearance package than anything else, it played upon the popularity of NASCAR and drivers, such as the late Dale Earnhardt.

Chevrolet Monte Carlo SS

Chevrolet's Monte Carlo SS was a styling success, but a performance failure. In my opinion, the Monte Carlo SS was one of the best-looking cars on the road during its 1983–1988 production span. It had an aggressive, body-colored front fascia and bumper, high-tech-looking (for the time) graphics, a rear spoiler, and the same 15x7 styled aluminum wheels used on late second-generation Z/28 Camaros. The interiors featured real gauges, instead of the "idiot lights" common at the time, and a tachometer.

The L-69 "High Output" 305 sounded good through its low-restriction dual exhaust, but power was underwhelming due to its meager 180 hp, even with the 3.73:1

This car retains the factory gauge cluster. However, many owners upgrade to an aftermarket or Monte Carlos SS gauge cluster. When installing a new powertrain, keep tabs on the health and performance of the engine. Automater, Sunpro, and others offer analog and digital gauges that provide this essential information.

A Chevy 502-ci big-block is installed in Kristi Bejarano's Monte Carlo SS; a Tremec TKO-600 handles the shifting duties. The car carries original paint, which means there is less chance that substantial body work has been performed. A repainted car may carry a lot of body filler, which means you may be faced with body work in the future. This has a nice cowl-induction hood, and a set of 16x8 GTA mesh wheels. (Photo Courtesy Rick and Kristi Bejarano)

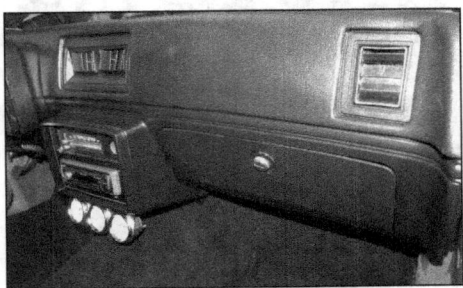

A refinished and capped dash with a plastic overlay hides the cracks in the original pad. This is a good, inexpensive way to make the dash look new. Like the door panel treatment, it is an effective aesthetic treatment.

This is the same Monte Carlo SS as above but it has been repainted white. This tone of white is much brighter than the original. (Photo Courtesy Rick and Kristi Bejarano)

CHAPTER 1

Rick Bejarano's Monte Carlo SS has had an engine swap, but it carries a Chevrolet Performance 502 big-block with Sanderson stainless-steel mid-length headers. While the Monte Carlo SS came with a TH-2004R overdrive automatic behind its L-69 305 HO engine, he replaced it with a 454 and a Muncie 4–speed manual transmission long ago. It now has a Tremec 5-speed to handle the 502's extra torque. It even has mechanical clutch linkage. (Photo Courtesy Rick and Kristi Bejarano)

gears that came in many SS models. It was still a carbureted small-block Chevrolet, though, and a lack of factory performance didn't stop many SS owners. Many received 350 swaps early on, some with Tuned Port Injection from the F- or Y-Body line, but the ultimate solution was as close as the nearest dealership parts department. The legendary LS6 and LS7 454 was still available (for about $2,500), and with few changes could be installed in the SS. Anyone attending Super Chevy shows in the 1980s may have mistakenly thought this was a factory option, as they were very common at the time.

The Monte Carlo SS has one of the best-looking factory gauge packages of the era, and the OEM gauges accurately display vital information. However, the factory gauges do not monitor all systems. On the A-pillar, an air/fuel ratio gauge monitors the mixture levels, and the boost gauge conveys how much air pressure is running through the engine. The Scanmaster 2.1 is a common accessory for 3.8-liter turbo Buicks. This scan tool keeps tabs on knock retard, oxygen sensors, and other aspects of the airflow and exhaust system. (Photo Courtesy Rick and Kristi Bejarano)

This beautiful Monte Carlo SS has been a drag car for most of its existence. It runs LS1 power.

Few G-Body cars carried a manual transmission, so a clutch pedal is a welcome, yet rare, sight. This pedal box in Bejarano's Monte Carlo SS is connected to a Tremec TKO-600 5-speed transmission. (Photo Courtesy Rick and Kristi Bejarano)

EVOLUTION NOT REVOLUTION

This Monte Carlo SS Aerocoupe was a rare model made for NASCAR homologation. Pontiac used a similar rear window treatment on the Grand Prix 2+2. The styling of the Aerocoupe didn't find much of an audience. Only 200 were made in 1986 and a similar small number was made in 1987.

This beautiful El Camino SS is a Holley Performance Products' development car. It features a 430-hp E-Rod LS3 engine from Chevrolet Performance with a T-56 6-speed manual transmission. The E-Rod engine delivers impressive performance, and it's emissions compliant in all 50 states, so you don't have to worry about licensing and making emissions limits.

Wheels-up is not an unusual state for a drag-oriented G-Body. This Monte Carlo LS has an LS1/T-56 combo swapped from a 2000 Camaro SS, and it's backed up by an early-style 10-bolt rear axle with 4.10:1 gears.

The Monte Carlo SS was very successful from a sales standpoint. NASCAR Cup Cars posted many victories around the country, and this on-track success spurred sales. The late Dale Earnhardt was largely responsible for the popularity, as he exemplified the "Win on Sunday, Sell on Monday" slogan that was common in the 1960s and 1970s, and had a brief resurgence in the 1980s.

The SS was successful, but the Ford Thunderbird driven by Dawsonville, Georgia, native Bill Elliott was more aerodynamic, and this made a big difference on high-banked tracks such as Talledega. General Motors' solution was to develop a fastback-style rear window, which was not only incorporated into the race cars, but sold as the Aerocoupe option, and as a retrofit kit through the parts department. These are the most sought after of Monte Carlo SS models today.

Drawing on the success of the Monte Carlo SS, Choo Choo Customs (now a part of Honest Charley's/Coker Tire) developed an SS package for the El Camino that was sold through dealers. The standard El Camino looked really dated due to the presence of chrome, which wasn't in style at the time. A styling package that looked similar to the Monte Carlo SS fixed that. This wasn't as simple as bolting on a Monte Carlo SS nose because the Monte SS is slightly wider than the Malibu. So, Choo Choo Customs built a new fiberglass nosepiece with similar styling, and a graphics package similar to the Monte SS graphics.

The El Camino SS appearance package is still available, so you can repair a real El Camino SS or build your own. It fits any standard Malibu, El Camino, Caballero, or wagon. Bulldawg Musclecars' project car, the *Grocery Getter* 1983 Malibu Wagon, was equipped with a Choo Choo Customs nosepiece.

Pontiac

Pontiac offered two G-Bodies in the early years of the body style. The LeMans was based on the Malibu, but with distinct Pontiac styling cues. These are rarely seen today, but are becoming highly sought after. There were two-door versions, but the four-door sedan and wagon are more prevalent today.

The Grand Prix was the slightly larger, more luxurious G-Body in the Pontiac line, and is not as popular as other models, such as the rare NASCAR-inspired 2+2 version. They are good-looking cars, but the grille is a love-it or hate-it item for most people.

Like the Monte Carlo SS Aerocoupe, the Grand Prix 2+2 was inspired by NASCAR competition, and has the same fastback-style rear window. Pontiac's existing Grand Prix nose was not very aerodynamic, and most agree it wasn't aesthetically pleasing, either. So, Pontiac designed a new nose for the 2+2 that was more aerodynamically efficient in the wind tunnel, but had styling that was and still is fairly controversial. You either love it or hate it. I fell into the latter crowd when they were new, but it has grown on me over the years.

The interiors were sportier than the typical Grand Prix, but the powerplant was the same lackluster 5.0 found in the Monte SS. Rare even when new, these cars are very collectible.

Oldsmobile

Oldsmobile offered its Cutlass models as the sole G-Body choice, but there were many choices within the Cutlass line. Two- and four-door sedans, as well as wagons, were available with plenty of options. Arguably one of the best looking of the G-Body line, the Cutlass Supreme two-door coupe was a very big seller for Oldsmobile. Most notable are the 4-4-2 and the even more rare Hurst/Olds 4-4-2 with its distinctive two-tone paint scheme and Hurst Lightning Rod multiple-handled floor shifter.

Oldsmobile 4-4-2

Oldsmobile's 4-4-2 had its own distinctive look and made the already beautiful Cutlass design even more attractive. The Olds Rallye

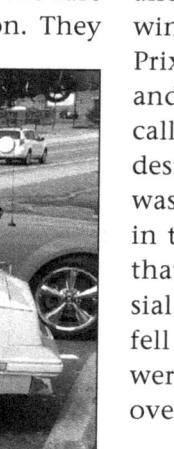

This Cutlass is in excellent original condition and is an ideal candidate for a project car. Even the vinyl top is in good shape, with no visible signs of rust underneath. Building a high-performance street, street/strip, or pro-touring car is often a challenging endeavor. If the car has substantial rust damage, you are taking on a massive project. A car restoration on top of a performance build is time consuming and expensive. Spending more for a car with a clean body and little rust is a wiser investment than starting with a rusty car requiring a lot of body replacement.

This Malibu coupe has a subtle performance look, sort of like a COPO car from the late 1960s. Equipped with a big-block and a 4-speed manual, this car is formidable competition for many street-going cars. G-Body cars are lighter than many F-Bodies, so big-block power with 4-speed transmissions makes them incredibly fast and fun to drive.

Scott Walkowiak's Buick Regal T-Type, known as the GNXray, carries the Turbo V-6, which was also used in the Grand National. This is one of the most prominent G-Body pro-touring machines in the country. Detroit Speed used this car as a test bed for its G-Body suspension parts. The front suspension features Detroit Speed shocks, sway bars, tubular control arms, front frame braces, lowering springs, and Track Kit camber shims. In the back it carries Detroit Speed tubular control arms, sway bar, and springs.

wheels bring back the essence of the early-1970s 4-4-2, and the striping adds just the right flavor; aggressive, but refined. Comfortable high-back bucket seats, full instrumentation, and a console with a floor shift harken back to the days when the first "4" was for 4-speed.

Unfortunately, it is a TH-2004R automatic. However, when combined with the typical 3.73:1-geared, 8.5-inch, 10-bolt rear, the cars always felt quick despite the lackluster Olds 307 under the hood. Hurst/Olds models came with the valet-proof Hurst Lightning Rod multiple shifters, which are a bit awkward, but cool looking.

Buick

Buick's G-Body offerings were limited to two-door coupes, four-door sedans, and four-door wagons, under the Regal nameplate. Known as more of a luxury brand, the Buick offerings were typically very well equipped with comfort options such as power seats, windows, door locks, air conditioning, and power steering. Surprisingly (or not, if you are a Buick GS fan), Buick was the first division to offer a performance-oriented model (the first Regal Grand National was released as a 1982 model), and most agree that the later 1986/1987 version was the pinnacle of not only G-Body performance, but of 1980s automotive performance as a whole.

From the start, hot rodders embraced the G-Body and started modifying them for increased performance. However, it took General Motors a while to recognize the performance market for the G-Body and offer any real performance models. The industry as a whole was concentrating on front-wheel-drive platforms with smaller four- and six-cylinder engines. Few realized the potential success for the G-Body. Other than a few notable examples, "performance" in the early to middle 1980s meant spoilers, tape stripes, and nicer wheels.

Buick Grand National

In 1982, Buick launched the Grand National, the first legitimate G-Body performance model. But the 1982 model was not very popular. It had a carbureted, naturally aspirated, 4.1-liter V-6 engine, and it didn't produce any more horsepower than many of the V-8s of the era. High-performance enthusiasts didn't accept the idea of a V-6–powered performance car. For 1984–1985,

The Buick Regal Grand National burst onto the scene in 1986 and became the fastest American production car. It was faster than the Corvette, Mustang, and all others on these shores. This immaculate example has a completely stock appearance, but has a built Stage 1 engine under the hood.

CHAPTER 1

A complete Buick Grand National drivetrain with 3.8-liter V-6 turbo and 200R4 automatic transmission is installed in this Monte Carlo SS, so this car turns heads when the hood is opened. Powertrains are often swapped across the G-Body platform, so don't assume a particular G-Body is all stock. (Photo Courtesy Rick and Kristi Bejarano)

In 1986 and 1987, Grand Nationals came with chrome-and-black 15x7 steel wheels.

Buick offered a 3.8-liter engine with a turbocharger and throttle-body fuel injection, which improved performance slightly. It also had a more sinister appearance, but these non-intercooled engines did not produce stellar performance. Known as the "hot air" turbo cars, they don't have near the potential of later versions.

In 1986, the Grand National took a huge leap forward in performance. The 3.8-liter engine featured an improved sequential electronic fuel injection (EFI) system, a coil-pack ignition, and an air-to-air intercooler that allowed increased boost levels. The Grand National was severely underrated at 235 hp, and 300 hp was more accurate. Off the showroom floor, these cars ran in the low 14s at around 100 mph in the quarter mile. Quarter-mile times of 12s were possible if you installed a performance chip, a less restrictive air intake, and a decent set of tires.

Soon, a cottage industry of Buick parts suppliers helped make the

Scott Walkowiak's T-Type, known as the GNXray, was featured in GM High-Tech Performance magazine. The car features Detroit Speed & Engineering suspension, front and rear Baer brakes, and CCW Classic wheels. The car is a rolling showcase for many parts offered by Walkowiak's company, GNS Performance. (Photo Courtesy Tony Wayman)

This T-Type has custom "GNXray" emblems.

12   GM G-BODY PERFORMANCE UPGRADES 1978–1987

"turbo Buicks" the most feared cars on the road. Buick engineers had officially built the car as a competitor to the Chevrolet Monte Carlo SS, but the Corvette was the real target. Buick trounced the Corvette, and had the fastest American production car in both 1986 and 1987.

Interestingly, the 3.8 sequential fuel injection (SFI) turbo engine was not restricted to the Grand National. Many loved the sinister look of the Grand National that featured a bulged hood, black paint and trim, and high-back bucket seats, but some traditional Buick customers were not fans of the styling. For them, Buick offered the Limited, with its Landau "receding hairline" vinyl top, pillow-back seats, and casket-handle door pulls, but it also had the same 235-hp (245-hp in 1987) turbocharged engine. The turbo engine was also available in

This 1987 Turbo T is roasting the hides. The car features 15x7-inch cast-aluminum wheels, which were used on 1986 T-Types as well as 1984 and 1985 Grand Nationals. (Photo Courtesy Rick and Kristi Bejarano)

The fender vents are exact reproductions of the GNX vents. Walkowiak sells them through GNS Performance.

Walkowiak built his own gauge cluster using AutoMeter Phantom-series gauges. The indicator LEDs tell him when the alcohol injection system is on and when the tank is getting low. This is critical to the survival of a turbocharged engine running higher boost pressure.

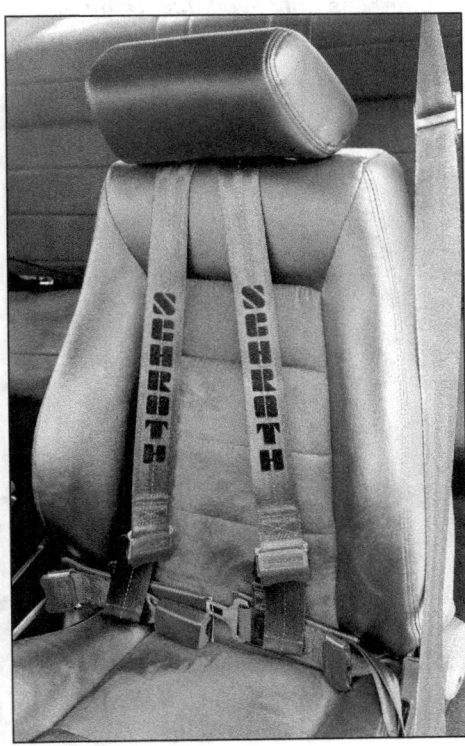
Instead of the typical light gray interior, this interior was reupholstered in leather and suede for a much nicer appearance.

the standard Regal, with any of the trim levels or colors available during those years, such as the T-Type in 1986 and the "Turbo T" in 1987.

A two-tone black and gray Designer Series T-Type was also available. By 1987, Buick couldn't build Grand Nationals fast enough, and most dealers were marking them up well over the sticker price (unheard of at the time). It didn't help that Buick had embarrassed GM's flagship sports car, and supposedly the number of Grand Nationals was limited. To get around this limit, and satisfy the demand, Buick offered the WE4 model. Other than different emblems and the Buick 6 logo conspicuously absent from the seats, it was essentially a Grand National.

As a final kick in the backside, Buick hired ASC/McClaren to build a limited edition, higher performance version of the Grand National, called the GNX. It had an improved turbo with a ceramic impeller, 16 x 8-inch cast-aluminum, mesh-style wheels (same as on the Pontiac Trans Am GTA), a unique torque-arm rear suspension, a chassis brace behind the rear seat, fender vents that recalled the old Buick "port holes," and flared wheel wells. Only 547 were built. This was the ultimate factory G-Body, and sold for $50,000, or more with dealer "market adjustments."

The 3.8-liter SFI turbocharged V-6 engine is known by the code LC-2. This engine was only available in the Grand National and T-Type in 1986, but it could be ordered in any Regal in 1987.

All Grand Nationals for 1986 and 1987 had the distinctive black-and-gray upholstery, a console with a floor shift, and a ridiculous gauge cluster. The speedometer only went to 85 mph, which was typical for the era, and the boost gauge and tachometer were small bar graphs. Most of these cars received aftermarket gauges, either in the existing dash (as on a GNX), or in a pillar pod (as on this one).

Many feared the Buick "Power 6" logo in the late 1980s, even Corvette owners. In 1986 and 1987, the turbo Buicks were the fastest American production cars. This engine, with a few slight changes, earned the title again in 1989 when it powered the Pontiac Turbo Trans Am.

CHAPTER 2

ON THE HUNT
FINDING YOUR IDEAL G-BODY

Vehicle selection is the most important decision you have to make for your G-Body project. While it seems very simple on the surface, many things have to be weighed and carefully considered before even starting your search. The keys to finishing the project are being realistic regarding every aspect of the buildup, formulating a plan, and sticking to that plan.

Choosing a Model

Which G-Body do you want? G-Bodies are versatile enough that any of them can be tailored to your preferred usage, build style, or choice of drivetrain, so it is more a question of brand preference and styling than anything else. Few G-Bodies came with real performance engines, and this is usually the first thing on the average enthusiast's list to change. If you want one of the performance variants, such as the Buick Grand National, Monte Carlo SS, 4-4-2, or Grand Prix 2+2, spend the extra money to buy the best real one you can afford, rather than trying to build a clone or rebuild one stripped of its special parts or badly damaged.

It is easy to fall into the trap of buying a lesser model or lesser quality car for cost reasons, then having a rude awakening when trying to find (and pay for) the necessary parts to replicate what you want. This is virtually impossible for the turbocharged Buick models (unless you have a nice donor), and difficult at best for the others.

Components, such as consoles, gauge clusters, bucket seats, floor shifters, spoilers, etc., can be very difficult to find. Beyond that, many are expensive. The restoration market has improved greatly for these cars over the past few years, but keep in mind that this isn't a 1969 Camaro or a 1957 Chevrolet so not many parts are offered as reproductions.

This 1987 Buick Regal Grand National looks stock, except for emblems that were deleted when the car was repainted. It has a race-prepped Stage II Buick V-6 for extra durability, a larger-than-stock turbocharger, and lots of bolt-ons. Owner David Huniker bought the car when new.

CHAPTER 2

These original appearing G-Bodies are great candidates for project cars. A Monte Carlo SS is shown on the left; a Cutlass is on the right. When purchasing a G-Body for a high-performance street, pro-touring, or street/strip buildup, buy the best car you can afford. Body panels, materials, fitment time, and paint make major body work an expensive endeavor. Therefore, if you buy a more expensive and structurally sound car, it saves you money over a cheaper and rougher car that requires extensive body work.

Chevrolet Malibus with solid floors, frames, and body panels are becoming more difficult to find. The car has always been popular with the drag race crowd, and now it is one of the most sought after starting points for a G-Body project.

Setting Performance Goals

Once you pick an appropriate model, the next step is to determine what you want from your project. I often talk to people who have purchased thousands of dollars worth of parts and spent countless hours making modifications, only to be frustrated because their vehicle doesn't perform the way they want it to. Sadly, spending a lot of money is no guarantee of quality or suitability for any particular purpose. Decide what you want out of the car up front, and work up a plan for achieving those goals.

The primary goal of this book is to provide solid and specific information on how to build your G-Body into the safe, reliable, high-performance vehicle that you want. When building a high-performance G-Body or repairing it, a GM shop manual provides excellent information and instruction.

Whether the car will be a daily driver, either during or after the buildup, greatly affects the choice of a starting point, as well as what modifications are suitable. Considering the age of even the newest G-Bodies, few are going to be reliable enough as a daily driver without a little work. Fortunately, these cars are very simple, and the most commonly worn mechanical parts are both inexpensive and readily available.

Emissions laws and/or state vehicle inspections need to be kept in mind, as well as gas mileage and occupant safety. That tunnel-rammed, big-block Chevrolet isn't the best choice for a 50-mile daily commute, and race buckets and a 12-point cage aren't going to work for taking the kids to daycare. That said, you can still have the performance you crave without breaking the bank.

ON THE HUNT

It is very tough to find a G-Body with an interior that is this nice, but they are out there. The custom gauge panel and dash insert feature Auto Meter gauges. The door panels are custom upholstered leather. It also carries the fourth-generation F-Body console, aftermarket seats, and roll bar.

This Malibu has a 408-ci LS engine with an 88-mm turbo for power and a full complement of aftermarket parts. The suspension, interior, body, and drivetrain have been customized. The fiberglass cowl-induction hood and the American Racing Torque Thrust II wheels are very popular additions.

Buying a rolling chassis is a good option if you plan to swap in a modern engine and transmission. A "roller" often sells for far less money than a complete running car, even if the body and interior are in excellent condition. In this case, owner Nick Freiser has cleaned and painted the engine compartment.

Creating a Budget

The next aspect to consider is one that most of us want to avoid: the budget. Most of us have a limited amount of money to spend on a given project. As a professional builder, the most difficult thing I do on a daily basis isn't building hot rods or running a business; it's coming up with an accurate estimate of the cost to do a given job.

Granted, I have to estimate labor as well as the parts costs, but even though most enthusiasts consider their labor "free," it still has a price, especially if you have a family and a demanding career or other vocation. With the many online resources currently available, it is easy to check prices without leaving home or picking up the telephone, but the little things that people rarely think about are the ones that really take a toll on the budget. Spray paint, bolts, hoses, fluids, lines and fittings seem insignificant on their own, but they all add up.

If you have never taken on a large automotive project, make your best estimate of the parts cost and,

When shopping for a G-Body, don't assume all the parts are stock. Most have been through many hands over their life span, and lots of changes may have been made along the way. In this case, it is a positive change, resulting in a fully baffled, home-built aluminum fuel tank.

CHAPTER 2

An LS-Powered Monte Carlo

At first glance, Jarod Lucas' 1987 Chevrolet Monte Carlo LS looks like a mildly modified stocker. Lots of original brightwork and the factory composite headlights accent the deep black paint. The package provides a welcome change to the ubiquitous Monte Carlo SS models. The cowl-induction hood is an obvious addition, but the only other exterior modifications are the Chevrolet Bow Tie added to the grille and the S10 ZQ8 cast-aluminum wheels.

Instead of the standard-issue 305-ci V-8 or 4.3-liter V-6, Lucas installed an LS1 backed by a T-56 6-speed manual transmission. After seeing the tall truck-style intake manifold, I thought he was running an aluminum 5.3 or 6.0 from a truck, but it is indeed an LS1, which he pulled (along with nearly everything else) from a 2000 Camaro SS. He prefers the torque characteristics of the truck intake over the passenger car LS1 or LS6 intake that most run, and judging by the car's performance I agree. This Monte Carlo was pulling the front wheels as it ran consistent 11.80s in the quarter mile at Beech Bend Raceway, in Bowling Green, Kentucky. The engine is stock, other than a custom ignition curve and bolt-ons, such as the intake, headers, and cold-air induction.

This isn't your typical LS swap, though. Lucas was very resourceful, using lots of parts from his donor Camaro. Rather than using just the engine and transmission, he swapped the radiator and fans, and he integrated the Camaro wiring harness into the Monte's harness. In addition, he retained all the Camaro's modern features, such as airbags, ABS, and the anti-theft system. He sourced the pedals for his T-56 swap and all the factory hydraulics from the Camaro. Also, the dash, seats, steering column, and console came from the Camaro. It took a lot of time to sort everything out and make it fit into the Monte but doing so significantly reduced his cash outlay. If you have the time and skills, this is a great way to go.

Jarod Lucas built and owns this 1987 Monte Carlo LS. At first glance, it appears to be a well-restored and mostly original car. However, a 320-hp LS1 provides the power and a T-56 6-speed manual transmits the power. Lucas pulled the powertrain from a 2000 Camaro SS donor car and used nearly everything to build the Monte Carlo.

The hood ornament has been shaved, and a Bow Tie from another model has been added. Lucas also installed the steel cowl-hood.

The 16x8-inch wheels were sourced from a late-model Chevy S10 ZQ8 truck.

The full interior from a 2000 Camaro SS has been fitted to this Monte Carlo. Even the airbags and factory anti-theft system are fully functional.

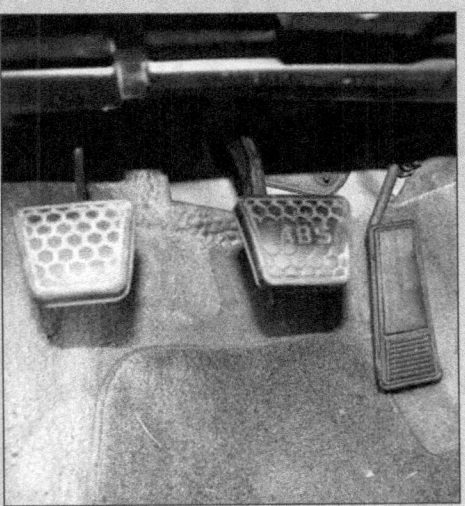
The pedal box from a 2000 Camaro SS was transplanted into this 1987 Monte Carlo.

This LS1 is fitted with a truck-style intake. This combination delivers increased torque available with this intake. The car runs 11.80s without any power adders, and is otherwise stock except for the headers and cold-air intake plumbing.

Using the entire wiring harness and fuse block from the donor vehicle is very resourceful.

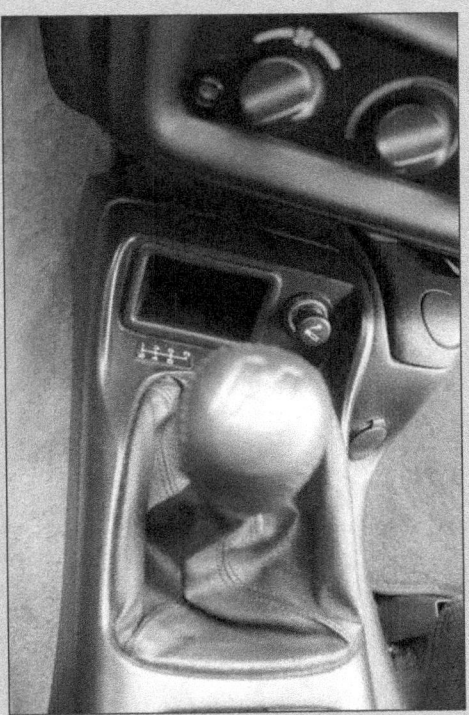
The console and Hurst shifter are stock Camaro SS. The T-56 6-speed manual transmission lurks underneath.

CHAPTER 2

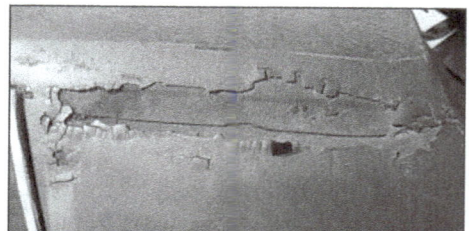

General Motors produced hundreds of thousands of G-Body wagons in the late 1970s and early 1980s, and because they are abundant, many are being transformed into high-performance cars. Some wagons have been turned into full pro-touring cars. This Malibu has been nicely reworked, with the bumpers tucked, custom paint, and aftermarket wheels. It's powered by a 5.3-liter truck engine and a 4L60E, which makes a great combination.

When subjected to high-horsepower loads and hard drag-strip launches, G-Bodies often cracked at the roof seam. Unlike most earlier cars, which used body solder (lead), plastic filler was applied at the factory to cover the roof seams. This leads to cracking and eventually to rust.

at the very least, double it. Triple the estimated time, and you are probably in the ballpark.

General Inspection

Once you have a budget, it is time to start looking for the right car. I'm not referring to just personal preference here, but the car that best fits your performance goals, budget, skill set, available time, and workspace. You have probably heard this before, but it still holds true: Buy the best car that you can afford.

A cheap, or even free, car that needs extensive metalwork to repair rust or collision damage is often no bargain, even if you are capable of doing the work. Plus, in many cases new parts just aren't available, especially body parts. The G-Body no longer has many "crash parts" available, and restoration-quality body panels are few and far between, especially if yours is a less popular model.

The Frame

Finding a solid body can be a challenge, particularly in the northern United States and Canada, but plenty of rust-free examples can be found in warmer climates, often at very reasonable prices. However, transporting a car cross country is still less expensive than even minor rust repair.

When inspecting a car, don't forget to look at the frame. G-Body frames were very weak, even when new, and minor accidents are often enough to seriously tweak them. Northern cars are also prone to frame rust, particularly in the rear, and even rust- and collision-free examples are often cracked due to flex. Several sources offer replacement frame sections for reasonable prices, but it's often more cost effective to just replace the frame with a better example if it has major damage or rust issues. Fortunately most G-Body frames interchange. The basic types have only minor differences in frame

This El Camino has fairly fresh paint; in many cases, that is an attempt to hide body problems. A lot of filler resides under the paint, especially where the fiberglass Lauren Engineering bumpers are molded to the (steel) body panels. The body filler presents problems and body work must be done. Body filler should be no more than 1/8-inch thick (preferably much thinner). A thick coat of body filler shrinks and eventually cracks, necessitating a re-do.

horns and engine and transmission mounting locations. The exception to this is that wagons and El Camino/Caballero models use their own specific chassis.

Exterior Paint

Paint and bodywork are often the most expensive parts of any automotive project. Although you may be able to do the paint work yourself, in many cases it doesn't make sense to do it yourself. It's very difficult to produce a professional-quality paint job at home because you must construct a temporary paint booth and keep contamination to an absolute minimum. Beyond that, a substantial investment in paint guns, compressors, other tools, and materials is needed. Therefore, if the car needs a paint job, find a professional shop to do the work.

Of course, you can try to find a car with a straight body and nice paint. Because most G-Bodies haven't attained collector car status (yet), finding one with a high-quality paint job may be a challenge, simply for economic reasons. A basic paint job with the jambs done, and with quality materials, can cost around $4,000; and that doesn't include stripping the car, or any priming or bodywork. If you have the capability to do it yourself, and a place to do it, you can save a fortune, but it takes time, tools, skill, and a facility to do it right.

Most local paint suppliers have a professional downdraft paint booth that can be rented for $100 to $200 per day, and often body shops rent their booths for a similar fee. Materials can easily run $1,000 to $1,500, or more, depending on brand.

If paint is less of a concern, or you don't want to invest that much time or money, a weekend of prep time removing trim, working out dents, filling, sanding, and spot priming gets your car ready for a paint job by one of the larger discount chains, such as Maaco, and you can put the money saved toward other areas.

Interior

The next most important area to consider when buying your G-Body is the interior. Some parts, such as the headliner and carpet, typically need to be replaced. If it wasn't replaced, the average enthusiast can complete the job relatively inexpensively. If the headliner board is in good shape, it can be reused; if not, most are still available through reproduction parts dealers.

Reproduction seat covers can be obtained for most popular performance models, either in the original

The cloth used in most G-Body interiors didn't hold up well in the sun. This Grand National seat back has the typical fading. Reproduction covers are available, even in leather (non-original).

A stained and worn original seat is common in many G-Bodies. The tear in this seat bottom looks worse than it is. The seat foam is still in good shape. Fortunately, replacement covers are available from several restoration suppliers, so these seats can be returned to like-new condition without complete restoration.

This 1987 Grand National (owned by Gary Tarwater) carries a professional paint job. It was stripped to the metal, primed, block sanded, and covered with basecoat/clearcoat. Factory paint was awful on most of these cars, so don't assume poor paint is the result of substandard repainting. You won't see a factory-painted car that looks as good as this one.

cloth, or leather/vinyl. Again, even a novice can recover seats with a little instruction and some patience, and no more elaborate equipment than a pair of hog-ring pliers and some side cutters. For bench-seat models, wagons, etc., you may have to enlist the services of a good upholstery shop. In many cases, original-style materials are still available, or more modern materials can be used.

Hard trim, such as the interior plastic quarter trim panels, A- and B-pillar moldings, and kick panels, are often faded, broken, or brittle. Fortunately, these parts interchange among various years and models, so finding good used replacements isn't that difficult, especially if you are more concerned with having functional replacements rather than finding identical parts for aesthetic reasons.

Interior trim paint is available in most popular colors in spray cans. For more unusual colors, and a bit more durability, any local automotive paint store can mix dyes and

This quarter trim panel is in excellent condition, just a little dirty. Chalky or discolored panels are easily dyed. Eastwood and many other companies offer spray-on dye to match interior colors. Eastwood has plastic adhesion promoter that helps dye bond to many plastic materials.

Unfortunately, some dash pads are beyond saving. This could be replaced, but what else was done to this Cutlass with a "Tropicana Orange Juice" theme.

This seat belt is in great shape, but the plastic cover is missing. You can find a cover from another GM car at the salvage yard; it can be re-dyed if necessary.

The tilt column is in great shape, but the trim ring around the horn button needs replacement. This is very common on Grand National and T-Type steering wheels.

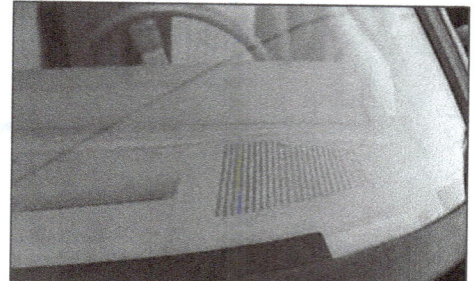
The dash pad is free of cracks, but the speaker grilles aren't. Fortunately, replacements aren't difficult to find.

paints in any color you want. They also typically carry all the prep products you need for the job, to ensure good adhesion.

Clean, used dash pads are a bit more difficult to find, especially for earlier models and cars from the Southwest. The later ones don't tend to crack as severely and are more widely available. Dash caps (molded ABS plastic covers that simulate the original soft cover) are available for most models.

New dash pads are virtually impossible to find, but good used ones can be found online, and in some cases they can be re-covered by a professional shop. Just Dashes, for example, can strip the old vinyl and foam, make any repairs to the structure that are needed, and reapply foam and vinyl in a vacuum-forming process. This is expensive, but is often the only real choice for restoring a hard-to-find dash pad.

Interior components specific to performance or sport models are difficult to replace if they are damaged or missing. Consoles are no longer available from General Motors, and none are currently being reproduced. Steering columns were specific to these cars (no shifter handle), and are often broken by thieves. If you at least have the original column, a specialist can often rebuild it.

Floor shifters are often broken or replaced with aftermarket versions. Fortunately most of the shifter linkages are similar to those found in other GM cars of the era and can be adapted.

Bucket seats can be sourced from another G-Body of the same type, but they are getting harder to find. While virtually identical seats were installed in other car lines, the seat brackets were different. For example, if you have a 1982–1983 Grand Prix with "PMD" buckets, and they are in bad shape, you can often find a better seat in a same year Firebird. The nicest set of PMD seats I ever found in a salvage yard were in, of all things, a 1982 Pontiac 6000 four-door station wagon. However, the seats did not recline because they were made for a four-door car. Fortunately, the hinge mechanisms from 1982–1992 Camaro or Firebird bolted right up, and once the G-Body seat brackets were attached no one could tell the difference.

The standard late–1980s Firebird seats are very similar to those found in Buick Grand Nationals and some T-Types, so many of the parts interchange. The tremendous interchangeability of GM parts also allows non-stock bucket seats from later models to be bolted in, as long as the correct G-Body seat brackets are used.

Sunroof/Moonroof

The factory-installed sunroofs and moonroofs on some G-Body models, particularly Regals, present inherent problems. Replacement parts are difficult, if not impossible, to find. They weaken the roof structure and often leak, which can lead to rust. Frankly, I suggest passing on any car that has a sunroof, regardless of condition, unless it is a factory moonroof in good working order.

As with T-tops, neglected sunroofs and moonroofs often leak, contributing to rusted floorpans, so be sure to check these cars especially well.

Electricals

Carefully inspect the electrical system. Look under the hood and examine the condition of the wiring harness. Verify that the connectors are properly installed and the harness is not worn, cracked, or chaffed. If there are a number of open terminals and loose wires, an amateur most likely did the electrical work and this can lead to a number of problems that can be far ranging. If too high of an amp load is running through a circuit, it can create a fire.

If the OEM radio is currently installed, it doesn't mean that it was installed for the entire life span of the car. Look for signs of previous stereo installations and other electrical repairs or modifications. Hacked sheet metal to fit larger speakers is common, as is shoddy electrical wiring and ruined dash bezels and interior panels. Similarly, activated and even deactivated alarm systems may

This Firebird seat is virtually identical to the seats found in most Regals, including Grand Nationals, and is far easier to find in a salvage yard. The seat mounts are different, however, and must be swapped with G-Body mounts.

CHAPTER 2

Scott Walkowiak owns this Buick Regal T-Type, which has a mostly stock engine compartment. Note that the troublesome Powermaster brake booster system has been replaced with a conventional one. Walkowiak was running a TE-44 turbo, ported heads, and a mild-performance cam at the time of this photo, for around 400 hp.

A typically modified G-Body engine compartment often looks like this. Nick Freiser owns this El Camino that has an Edelbrock intake manifold, carburetor, and headers, but is otherwise stock. This shows what a little cleaning, repainting, and the installation of a few dress-up parts can do.

cause shorts or problems in starting the vehicle due to poor installation.

HVAC System

Another area that is difficult or expensive to repair is the HVAC system. The heat portion is rarely a problem, but it can be rendered inefficient or even inoperable by missing or deteriorated vacuum lines or a plugged or leaking heater core. If the car has a sweet smell, this indicates that some antifreeze is present in the interior. It may be leaking, so the floorpan on the passenger side should be thoroughly checked.

Closely inspect the air conditioning system. If the belt was removed from the compressor, the clutch may have locked up and the belt has simply been removed rather than getting an expensive repair. The brackets are difficult to find if missing.

If the owner says that the A/C system "just needs charging," it most likely has other issues

The casing for the A/C core on the firewall is often damaged, especially the lower parts, but in most cases, minor cracks and broken areas can be repaired.

If the whole system was removed and the firewall blocked off, find another car if you want HVAC, or adapt a modern aftermarket system to the car. It would probably be cheaper and easier in the long run and work far better than the stock system.

Emissions

Strict emissions laws and vehicle inspections aren't relevant to everyone. It is a concern in regions where you have to worry about the presence and functionality of EGR valves, catalytic convertors, and other emissions components on 1986-and-older vehicles.

If contemplating a project in a state with strict regulations, such as California or New Jersey, I recommend looking closely at the laws,

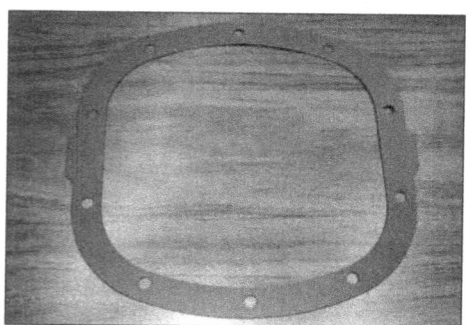

This is the gasket for a 7.5-inch 10-bolt rear end, which has been found in most RWD GM cars since the Vega. Although inherently weak, they are sufficient for most street applications. The bolt pattern is the easiest way to identify it.

The 8.5-inch 10-bolt rear end is very strong. They are commonly found in turbocharged Buick models and some Cutlass 4-4-2s. While it looks very similar to the 7.5-inch, the cover and the bolt pattern are more circular, rather than the slightly square 7.5-inch.

and whether any exemptions are available for your planned modifications.

Also keep in mind that tampering with or removing federally mandated emissions equipment such as smog pumps or catalytic convertors is illegal, regardless of local laws.

Engine and Drivetrain

The condition of the engine and drivetrain components must also be evaluated. A project car that will be a daily driver needs to be mechanically sound. Even if you intend to gut the car and replace everything a thorough inspection gives you time to find any issues and negotiate a lower price.

Closely examine the engine for fluid leaks, missing or altered components, odd smells, and any obvious signs of repairs or modifications. Also check the fluid levels, and pay attention to the condition of the fluids. A burned smell in the oil or transmission fluid or an excessive fuel smell in the oil can point to other problems. Fresh fluids are often a telltale indicator about how the car was maintained. A car that has a low coolant level in the radiator or is running black oil indicates a lack of regular and necessary maintenance.

If originality of the engine is important to you (especially with turbo Buick models), you need to decode the ID tags. (You can find this infomation on the Internet.) This is also a good idea for the transmission and rear end, as I have seen many examples of cars that came with more desirable valve bodies in the transmission (TH-2004R) and upgraded rear ends (8.5-inch 10 bolts, or 7.5-inch 10 bolts with "performance" 3.42 or 3.73:1 gears) swapped out due to breakage, theft, or outright fraud. This can also work to your advantage, as 8.5-inch 10 bolts or earlier A-Body 10- or 12-bolt rear ends (all stronger than the standard 7.5 rear found in most of these cars) are often swapped in, especially in cars that have seen drag racing.

After performing a thorough visual inspection, take the car on a test drive. This is where you can drive the car at varying speeds and conditions to uncover any significant operational problems that are not obvious, such as slipping transmissions, whining rear ends, odd handling or braking traits that can point to past collision damage or needed repairs, etc. Even if you plan to swap in that LS3 and T-56 as soon as you get it home, this can give you more ammunition when negotiating a price with the seller.

In most cases, G-Body cars already have an entirely different engine and transmission than they came with from the factory. The vast majority of G-Body cars came with uninspiring 3.8- or 4.3-liter V-6s, or small V-8s such as the Olds 260 and 307 or the much-maligned Chevy 262, 267, or 305 engines. A few early models had the Pontiac 301. None of these were very powerful engines, even for their time, and the vast majority of those still running have already had the engine replaced.

Still, if you are going with the existing engine, have all the proper mounts, brackets, and accessories so you can save time and money with the engine, even if the engine itself is going to be replaced. If you plan to retain a small-block Chevrolet of any displacement, the existing engine mounts and accessory brackets can be used. The same is true in most cases for Olds engines, which many G-Bodies came with. Whether you have a 260 or 307, a larger "Rocket" 350, 400, or even a 455 bolts up with very little modification. For Pontiac engines, the 301 mounts can be used, or those from another, earlier, Pontiac can be easily adapted.

If you plan to swap a Buick or Cadillac engine into a G-Body, or swap in a later LS-series engine or big-block Chevrolet, you may be using everything from the donor vehicle (likely a full-size sedan, or an older A-Body), or using aftermarket parts, so condition and completeness isn't an issue. A "roller" or an otherwise complete car without the engine and transmission is ideal for this kind of project.

Rust Issues

Rust can be a very serious issue with almost any car, and the G-Body is no exception. Many do not have a chance to buy a car from a rust-free environment such as Arizona, California, and Texas. Even if the car is currently in a rust-free environment, it may not have spent its entire life there.

In its early stages, body rust is often visible as small bubbles under the paint, particularly on the vehicle's lower surfaces. This is most evident in the bottoms of the front fenders, the front corners of the doors, and the lower rearmost parts of the quarter panels. Cars that had wide trim, such as the Regal Limited, and a vinyl top should be scrutinized very carefully.

If the vehicle has its original paint, any rust is more obvious, and you are able to properly evaluate the condition of the body. That said, be wary of freshly painted cars for sale, and particularly any "just needs paint" deals. A car that "just needs paint" usually

CHAPTER 2

Be wary of cars with vinyl tops. This is a fairly extreme example for a Southern car, and would be very difficult to properly repair. I do not recommend buying a car with a vinyl top, unless you plan to remove it permanently. General Motors did very little, if any, rustproofing in this area.

The windshield header is another area to watch for rust, especially on T-top cars like this one.

This El Camino has extensive rust damage, most notable here in the bed. Unless you have good metal fabrication skills or a suitable donor, I would pass on this El Camino.

means "I filled all the rust and dents full of body filler, then rattle-canned the car in gray primer and left it outside in the rain for a year."

Always bring a magnet when inspecting a car. Place the magnet at strategic positions on the body, such as the bottom of quarter panels, wheel well lip of the front fender, rocker panels, and other areas. Be sure to inspect the trunk (especially under mats and carpeting), and that includes reaching your hand into the quarter panels. If it feels especially rough, or crunches, you have heavy rust in the quarter panels.

Also look for "worms" in the backside of the panel. They are the evidence of a body shop pulling dents with a drill and slide hammer. It may look okay now, but eventually leads to rust as moisture gets through the filler to the unpainted steel below it. If possible, pull up the sill plates and peel back the carpet to inspect the floorpans. If that isn't possible,

You are looking at a substantially rusted floorpan. This car does not have a T-top or sunroof, so it is otherwise rust free. You can examine the condition of the floorpan by pulling back the carpet, or better yet, examine it from underneath. Parking in grass, especially tall grass, for extended periods often causes rust like this. Leaking heater cores that aren't repaired promptly can contribute to the front passenger pan rusting.

A good candidate for a project, this El Camino is mostly original and has only minor damage to the door. The School of Automotive Machinists competed with this car in Holley's Engine Swap Challenge.

When inspecting a G-Body, don't be bashful. Crawl under the car and look at the floorboards, frame rails, and trunk. Identify corrosion and crash damage that needs to be fixed. Use a magnet and place it on the fenders and quarter panels, particularly around the wheel well lips. If the magnet does not stick, you know there is a lot of body filler under the paint, which indicates future problems.

This clean Monte Carlo LS truly needs nothing. The paint and body are excellent, the interior has been replaced with one from a 2000 Camaro SS (including the dash and all wiring), and it has LS1 power. It also has a T-56, which is becoming more common in G-Body cars.

This early G-Body Cutlass has seen better days, but it is restorable. No rust or major body issues make this a perfect project candidate. Finding a new grille insert isn't too hard; you just look for a car that's in similar condition. On the other hand, rust on any G-Body equates to major time and money in cutting it out and replacing body panels. For many common G-Body cars, it's not worth investing the time or money.

You should pass on cars with major body damage. G-Bodies are too plentiful and it is too expensive to fix this kind of major damage. In this case, the wheel came off and destroyed the quarter panel. I didn't take my own advice, however. This 1983 Regal T-Type is my own, and is being transformed into the GNXcess.

Worn-out door-hinge bushings often damage the striker. As a result, the doorjamb contains cracked metal. This issue is prevalent on G-Bodies because they carry heavy doors but thin metal in the doorjambs.

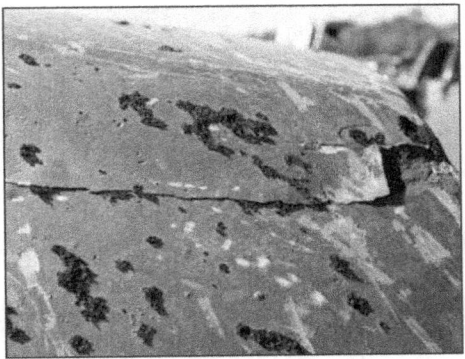

The roof seam between the quarter panel and the roof is cracked and rusted. G-Body coupes without T-tops typically crack at the forward part of this seam, especially if they are high-horsepower cars. Reinforce this area, especially if you're repainting the car. I recommend tack welding a solid metal strip in this area and grinding it smooth to prevent any future cracking.

get under the car to inspect the floorpans. You may see signs of a previous owner trying to hide rust, such as using spray-on undercoating or even roofing tar.

Carefully inspect models with T-tops or sunroofs/moonroofs because they tend to leak. Rust in the channels is actually more common than a failure of the gasket itself and it doesn't take long for floorpan rust to occur. You can easily fix the leaks with new weatherstrip.

Body Damage

Carefully examine the doorjamb area, especially the metal around the door striker. Due to the weight of the doors, the hinge pins and bushings often wear out, leaving a sagging door. If not corrected, it can cause cracks and metal distortion in the jamb around the striker. I have seen strikers that were heavily worn due to this, and in some cases, the striker itself becomes loose in the jamb, or is ripped from the metal. This is repairable, but if a car you're purchasing is suffering from this problem, it will cost you to fix it. Replacing the pins and bushings themselves isn't a big deal, but is most easily done with the doors off the car (which requires removing the front bumper and fenders), so take that into consideration.

Body panel alignment isn't as good as on some newer vehicles, but consistency is what to look for. Too wide (or too narrow) body gaps can often be the result of collision damage that wasn't properly repaired. A crack often develops in the B-pillar area (coupes only), where the pillar meets the roof, just behind the drip rail. This is particularly common on high-horsepower cars, such

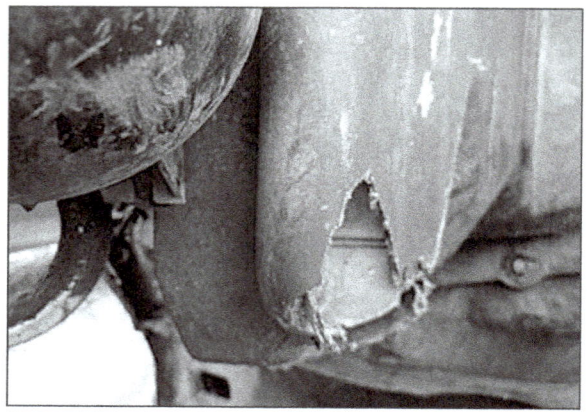

HVAC boxes are made of a composite material, and are often cracked or have pieces broken away. Fortunately, this particular one is easy to repair, but changing out an entire system requires a lot of work. If you want to retain the factory A/C system, find a unit with as little damage as possible. In the case of the GNXcess, we are building a new cowl/firewall section to make room for a Vintage Air system under the dash. This cleans up the firewall considerably. In addition, it makes servicing and packaging the twin turbos much easier.

as T-Types and Grand Nationals, and oddly enough, it is most prevalent on hardtop models.

You should repair it by carefully grinding then welding the crack, but it obviously isn't something you want to do on a car with fresh paint. The GNX came with a brace behind the rear seat (you can install a reproduction part on any G-Body coupe) to help correct this issue.

Finally, note any missing or damaged exterior trim pieces. Most of them are not available new.

Chassis Considerations

General Motors equipped G-Bodies with an inherently weak chassis. The geometry is good and it has adequate handling for most purposes, even in stock form. However, the frame itself is mostly open C-channel, and it serves mostly to locate the suspension and bumpers rather than for any real stiffening. It seems that GM suspension engineers relied on this, as much as any real suspension tuning, to deliver a smooth ride.

In stock form, this chassis was adequate for mild street use, but it certainly does not provide acceptable torsional rigidity and stiffness for a high-performance street build or a pro-touring car. After more than 25 years of use and abuse, many have suffered fatigue and rust corrosion. As a result, most of them need repair and stiffening. The open-channel design collects water, salt, and mud in the center, which leads to rust. The rear sections are also vulnerable to this kind of damage. Frame rust is an area of concern especially for northern climates where the roads are salted in winter.

Rough roads often led to frame cracking, especially in the rear sections, which exacerbates rust issues. Cracks are often evident around the steering box area as well. I recommend inspecting the chassis fully for cracks, rust, and fatigue as well as missing body mounts.

Excessive flex and age often deteriorate the rubber bushings and, particularly, suspension bushings. All suspension bushings, as well as the ball joints, should be inspected as well. If they have never been replaced, this would be a good first project since it is safety related, and also greatly affects handling and ride quality.

CHAPTER 3

MAXIMIZING STOPPING POWER
STOCK AND AFTERMARKET BRAKES

G-Bodies are fairly lightweight, especially compared to the previous A-Body, and as a result General Motors determined larger brakes were not needed. All models were of a power-assisted front disc/rear drum arrangement, with very little difference among models. Differences were in the type of power assist and, in some cases, rear drum material.

Front brakes on the G-Body are very conventional, readily available, and very inexpensive. A 10.4-inch, cast-iron, one-piece rotor and hub assembly was installed on G-Body cars, which were similar to those used on other GM vehicles since 1967. The metric, or quick take-up, caliper is a conventional, iron, single-piston caliper that features a low-drag design built into the brake system, which reduces friction and, therefore, increases mileage. Like the second-generation F-Body, the G-Body used a spindle with a cast-in caliper bracket, rather than a bolt-on one as found on most other models at that time. The steering arm is also cast into the spindle, as on the F-Body.

Rear brakes on all models have finned drums, which provides additional cooling. Most of these drums are iron. Some G-Bodies came with aluminum drums, but this was most common on Buick T-Types. Oddly, the Grand National received iron drums. The drums are completely interchangeable with one another, with no modifications required.

All G-Bodies had power assist, and the vast majority used a single- or dual-chambered vacuum booster. In addition, they had an aluminum-bodied, dual-reservoir master cylinder with a plastic reservoir and top cover.

The vacuum booster and master cylinder on the majority of G-Body cars look like this. Note that an adjustable proportioning valve has been spliced in to better control lockup of the rear discs. The kit is available from GNS Performance.

CHAPTER 3

By today's standards, these stock front brakes are adequate at best, even for a stock vehicle. A typical G-Body stock rotor measures 10.4 inches in diameter. The single-piston caliper is a "metric" type that's also used on third-generation F-Bodies and S10 trucks, so parts availability and interchangeability are very good.

All G-Bodies came with finned rear drum brakes. The only exception is the Regal T-Type, which came with aluminum drums that are otherwise identical, aside from saving a little weight. This iron drum was installed on a Grand National, but the design is no different than other models.

The earliest models (technically, still A-Bodies) used a conventional, iron, dual-reservoir master cylinder, but all later models used a quick take-up aluminum master cylinder.

Hydroboost Brake System

General Motors installed a hydraulically assisted unit called Hydroboost on some early Olds Diesel models. Pre–1986 turbocharged models, such as the Grand National, received it. Since 2007, Bosch has owned the trademark and calls it HYDRO-BOOST.

Installing Hydroboost is one of the most popular systems for increasing braking performance and brake pedal feel on a G-Body and many other vehicles. The GM Hydroboost power brake system was originally designed for low-vacuum vehicles, such as Diesels and turbocharged vehicles. Rather than using a vacuum diaphragm booster, the brake assist comes from the power steering system. The Hydroboost system doesn't require a special power steering pump; any stock GM pump is sufficient. Since it doesn't require engine vacuum, it is ideal for supercharged or turbocharged

Suitable donor units, such as this one, are found on the Chevy Astro and GMC Safari vans, but you have to visit the salvage yard to get one. If the pedal is hard and there are no obvious signs of leaks or corrosion, the brake system is suitable for a swap. If there are signs of wear or damage, rebuild kits are available from any parts store.

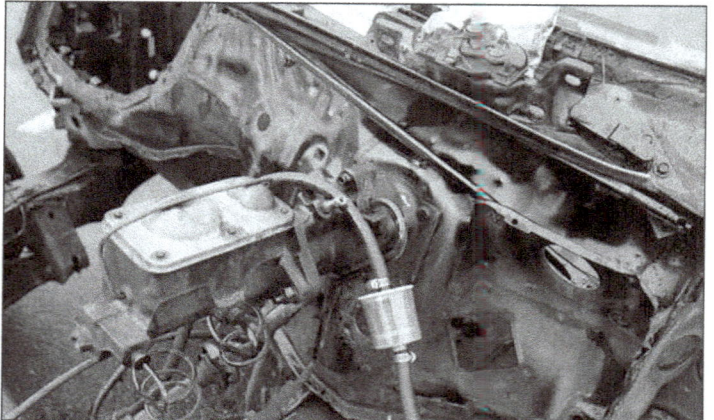

Hydroboost was installed on the 1983–1985 "hot air" (or non-intercooled) turbocharged Buick Regal. Instead of relying on vacuum, the Hydroboost uses hydraulic pressure from the power steering system. These units have become very popular as an upgrade for other cars. As a result, G-Body units are relatively scarce. In addition, some feel they don't work as well as later units. Remanufactured ones are available through most parts stores.

Hydratech offers this aftermarket version of the Hydroboost, using a late-model Corvette master cylinder and a Baer proportioning valve. A complete system retails for about $1,000.

applications. It also is more compact than a vacuum booster, which is a concern with some engine swaps, especially with big-block engines or engines running tall valve covers for valvetrain clearance.

Hydroboost came on pre–1986 turbocharged models, such as the Regal T-Type, Grand National, and some Diesel B-Bodies. They are a bolt-on swap to any other G-Body but it's difficult to find complete systems in good condition.

Another common source for Hydroboost swaps is the Chevy Astro/GMC Safari. This unit is easily adapted to the G-Body but requires custom-length hoses from a hydraulic shop, and some fabrication/modification of the unit's pushrod and mounting plate. You can use it with the factory G-Body master cylinder, as the stepped Astro unit is very odd looking and has limited fluid capacity.

Hydroboost is plumbed into the power steering system and uses very high pressures rather than vacuum from a conventional booster. This works well in low-vacuum applications, and is used in many full-size trucks as well as in some vans. These units have become popular for retrofits, and companies such as Hydratech sell these reworked GM systems.

Powermaster Brake System

The Powermaster system was installed on 1986–1987 turbocharged Buick models (Grand National, T-Type, WE4, Limited, Regal, etc.). It used a built-in electric pump to pressurize the system and a small, round reservoir (the accumulator) to store the pressure. The Powermaster was problematic, however, and initially very expensive to replace. It is rare to see one still in service now, except on low-mileage, original cars. Many cars that were originally equipped with this unit were switched to conventional vacuum boosters because they work much better, are inexpensive, and are a straight bolt-on modification. The Powermaster was also used on some Chevrolet Caprice models, notably wagons.

As most Grand National owners tell you, the stock Powermaster brake system performed very poorly. It suffered from problems such as bad accumulators and vacuum pumps that either quit or never shut off and gave lousy pedal feel. Many owners installed a Hydroboost that provided greatly improved braking performance, but owners often upgraded to a conventional vacuum booster and master cylinder.

To upgrade, you need a G-Body brake booster with (perferably) a dual diaphragm, new master cylinder, 2 to 3 feet of vacuum hose, and some vacuum tees. Also get a new vacuum manifold designed for the upgrade. The manifolds used on the 1989 Turbo TA, available from Buick parts vendors, are good units to source. You also need a brake pedal from any GM G-Body with a conventional

The Powermaster brake system installed on 1986–1987 Regal turbocharged models is often discarded for vacuum booster systems. I do not recommend using this unit for anything other than an original Buick turbo car. These problematic units deliver inferior brake performance compared to a conventional booster or a Hydroboost brake system. The accumulators often need replacement and the electric motors are known to stop working.

booster. This is a very simple upgrade that bolts on, and gives any turbo Buick a much better brake system.

Brake Upgrades

When contemplating a brake upgrade, several factors need to be considered.

Vehicle Use

Is your car going to be a daily driver and occasional drag racer? Are you building it for autocross events or "track days" at your local road racing venue? Just building a nice cruiser? Determine the application of the car before you buy any components because changes later on are expensive.

Cost

Cost is the next thing to consider. You can rebuild a stock system using all new parts for just a few hundred dollars, while installing a full aftermarket four-wheel-disc system, with six-piston calipers and huge rotors, typically costs $2,500 or more. Your brake rotors and calipers must be compatible with your wheels (i.e., the wheels must be large enough to fit over the brakes).

Wheel Size

If you aren't sure if your existing wheels, or the ones you want to use, will fit with your choice of brakes, be sure to ask about it. Any brake manufacturer can usually provide a template and tell you for sure which wheels fit with the new brakes. A good dealer can also help you determine the compatible sizes of brakes and wheels.

Most people have at least a rough idea what diameter wheels they want to run ahead of time, so this is a logical place to start. If you plan to use 15-inch wheels with an OEM-type single-piston iron caliper, choices are limited to stock or close-to-stock brake diameters.

On the other hand, if you decide to use a larger wheel diameter, such as 18, 19, or 20 inches, almost all available aftermarket rotors are compatible. Caliper clearance must still be kept in mind, though, as even OEM-style multi-piston calipers often interfere with common wheel styles. Pay particular attention to this if you are using spoke wheels, such as an American Racing Torque Thrust II. Many brake upgrades require a wheel spacer or adapter to use these wheels.

For budgetary reasons, and a desire to keep the existing wheels, many choose to stay with an essentially stock braking system. While a stock system does not give the braking performance expected from a modern car, it can still be improved upon slightly.

Rotor Type

High-performance cross-drilled and slotted brake rotors are a typical upgrade, but in a street application they have absolutely no measurable benefit. Inspect the existing rotors and make sure they are within spec (both for thickness and runout); if they aren't, replace them with new rotors. Repack the wheel bearings or replace them, as needed, while you have the rotors off.

As long as the calipers aren't leaking or exhibiting uneven pad wear, there is no need to replace them. Just be sure the pins are well lubricated with lithium grease (but don't apply too much!), and put in a fresh set of high-quality pads that meet your driving style.

Brake Hoses

Carefully inspect the brake hoses (both front and rear) and replace them if they look original or show any signs of cracking or swelling. DOT-legal, stainless-steel braided brake hoses are a good choice because they don't expand as much as standard rubber lines, and give a slight improvement to pedal feel.

Rear Drum

For the rear brakes, a set of aluminum drums is an optional upgrade, but only if you are concerned with keeping the weight down. Aluminum drums came on few G-Body cars (mostly Buick T-Types, but oddly enough not Grand Nationals), but are quite common on third-generation Camaros and Firebirds, particularly V-6 models. I am not aware of a source for new aluminum drums (the replacement ones are iron), but since these rarely wear out, used ones that are still in spec are pretty easy to find.

Brake Shoes and Wheel Cylinders

Inspect the brake shoes and replace them if needed. Again, look for uneven wear, and any signs of brake fluid on the linings or hardware. This is a sign of a leaking wheel cylinder, and must be replaced immediately. If one side is leaking, the other side will probably be leaking soon as well, so replace them in pairs. If the G-Body is intended for the drag strip, you may want to consider two modifications often made by turbo Buick owners. One is to replace the wheel cylinders with ones from a late-1980s S10 truck, which are larger and allow more fluid volume to help lock the rear brakes when launching. This also gives slightly better performance from the rear brakes.

Selecting Brake Fluid

Using a high-quality brake fluid provides the largest improvement in a stock brake system. When selecting a brake fluid for a mostly street-driven car with stock brakes, I typically use Ford heavy-duty DOT 4 brake fluid, which is readily available at any Ford dealer. For a modified braking system, especially one that sees track use, I prefer Wilwood 570 brake fluid.

Properly flushing the brake system and bleeding the brakes also gives a dramatic improvement in pedal feel and braking performance. Brake fluid is hygroscopic, meaning it absorbs moisture. If moisture is present in the lines, the pedal can feel spongy, as well as damage brake system components through rust and corrosion.

To flush the old, contaminated fluid, first remove as much fluid as possible from the master cylinder reservoir. Use a turkey baster or soak up the old fluid with clean disposable shop towels. Then fill the reservoir with fresh fluid. Once the reservoir is full, bleed the brakes as normal.

Several methods can be used to bleed the brakes, from the traditional two-person method to modern pressure and vacuum bleeders. The method you use is up to you. I generally use the Phoenix Injector, in either RFI (reverse fluid injector) mode or vacuum mode. This is a great tool to have, especially if you are working alone. The Phoenix Injector is available directly from the manufacturer or from retail outlets such as Summit Racing.

For many, the stock braking system just doesn't offer enough performance. Fortunately, however, there are plenty of solutions to this problem, including those that primarily use OEM components. This type of upgrade has several advantages, including OEM reliability, parts availability, and relatively low cost compared to aftermarket solutions.

The second modification is to purchase two sets of brake shoes, and use only the primary shoes (the ones with longer friction material). This gives a little more friction surface, and further helps to lock the rear brakes. Some turbo Buick specialists sell brake shoe kits with only primary shoes, but typically there isn't any cost savings over buying two sets of conventional shoes.

Spindles

For many years, the most popular brake upgrade for G-Body enthusiasts was an upgrade to full-size GM B-Body spindles. These are now a little harder to find in salvage yards. The full-size B-Body spindle is "tall," similar to the second-generation F-Body (Camaro/Firebird) spindle, and in its original application works very well. The upper ball joint mounting point

The 1998–2002 S10 Blazer 2WD spindle is a popular upgrade for G-Body cars. This aftermarket spindle from Belltech features a 2-inch drop and is similar to the S10 spindle. Most rotor sizes can be bolted to this extremely versatile spindle because of the bolt-on caliper bracket design. The original spindles, with their cast-in caliper bracket, limit the rotor size.

Unlike the stock G-Body spindle, this Belltech spindle uses a bolt-on hub with sealed bearings for improved performance and reliability.

is higher than on the stock G-Body spindle, so it changes the geometry and produces negative camber. This spindle uses a 12-inch rotor with an integral hub, and an iron single-piston caliper that is slightly larger than the stock one.

On the surface, this seems a good swap, but in reality it introduces several problems. For one, the stock upper control arm on a G-Body doesn't allow enough adjustment, even with offset control arm shafts and a large shim pack. This means that shorter aftermarket upper control arms with altered ball joint mounting points designed for the taller spindle are mandatory.

Many enthusiasts use aftermarket upper control arms, some of which are designed for a tall spindle anyway, so that in itself isn't reason enough not to use them, but the added expense needs to be considered. If the cost of properly doing this swap were the only issue, it might be worth doing (or retaining). However, using the tall B-Body spindles creates other problems.

For instance, the steering arm is cast as part of the spindle, similar to stock G-Body and S10 spindles. But the arms are longer and at the wrong angle for a G-Body, so the results are massive bump steer, a slower steering ratio, and a negative effect on the Ackermann geometry.

The factory-style B-Body rotor has a 5-on-5-inch bolt pattern, as opposed to the standard GM 5-on-4¾-inch, so it is necessary to either re-drill the rotors or use an expensive and hard-to-find 1LE rotor from a third-generation F-Body. The single-piston Delco caliper is larger than the stock G-Body caliper, but it is also heavy, a drawback of this whole system.

If you want to use a larger rotor/caliper combination on this tall spindle, it can be done, but as with the stock G-Body spindle it requires machining the stock caliper bracket off and using a custom caliper bracket for the desired caliper and rotor combination.

Although tall spindles are a nice upgrade, they are not really necessary.

Spindles from Blazers and Jimmies

Installing the spindles from 1998–2002 S-series 2WD Blazers and Jimmies on G-Body cars is an effective handling upgrade and an excellent basis for future upgrades. These spindles don't have a cast-in bracket for the caliper, and they also use a separate sealed hub. You can use the factory GM spindles, which are easily found at a salvage yard. "Dropped" spindles are available from Belltech. However, a dropped spindle does not change the suspension or the steering geometry in any way. It is a "short" spindle like the stock design.

G-Body enthusiasts have long taken advantage of the parts interchangeability between the Chevrolet/GMC line of S-series trucks, for aftermarket parts, such as dropped spindles, front coil springs, etc. However, the greatest contribution you can make to a G-Body from the S-series parts bin is to use spindles from 1998–2002 S-series 2WD Blazers and Jimmies. The spindle is dimensionally identical. It bolts up

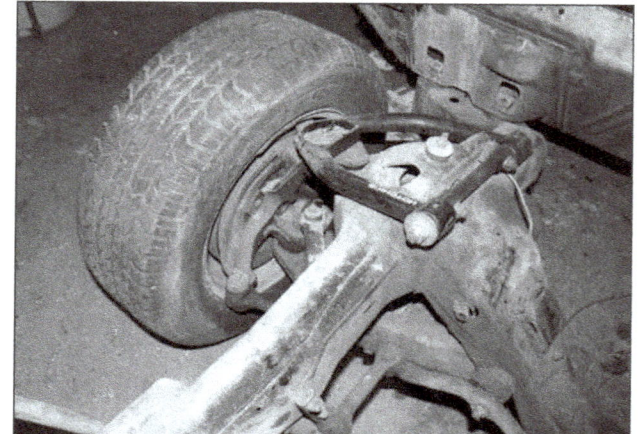

This is the "old school" way of getting larger discs on a G-Body: the B-Body spindle and caliper. This does give a taller spindle, which is desirable, but it creates a really bad angle on the tie rods and requires a longer upper A-arm. The one pictured is installed on an early A-Body.

These rotors and calipers are compatible with G-Bodies. On the left is the stock brake rotor and caliper. In the middle is a C6 Corvette Z51/J52 brake rotor with a C5 caliper. On the right is a larger Grand Sport rotor. You could use Z06 or ZR1 parts if desired. The 1998–2002 S10 Blazer 2WD brake rotor (not pictured) uses a slightly larger two-piston caliper.

MAXIMIZING STOPPING POWER

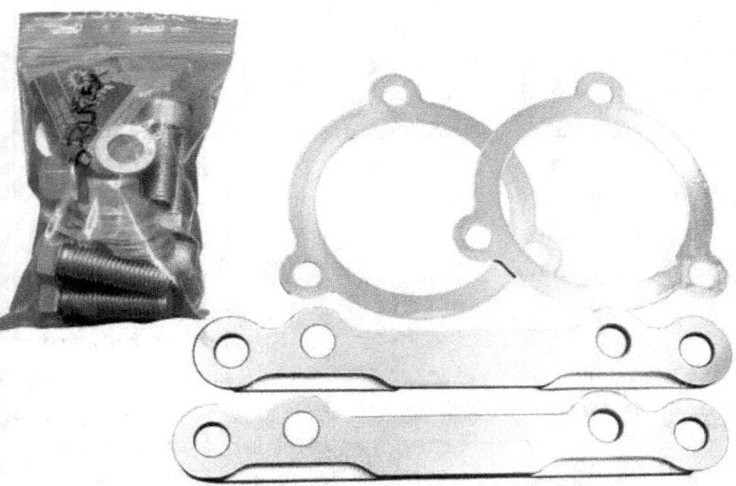

This KORE3 bracket kit replaces the stock caliper bracket. It allows any C5 or C6 caliper abutment to be bolted on, so you don't have to cut off the stock caliper bracket and bolt a new one to the original spindle. The kit includes all the necessary hardware, shims, and thread-locking compound.

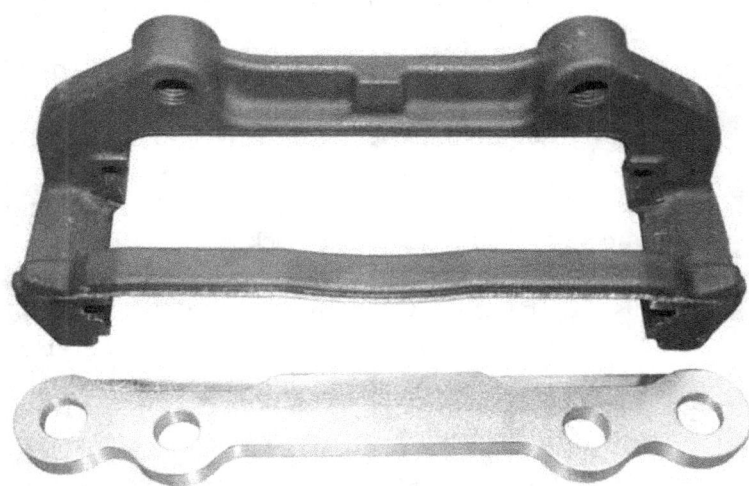

A C6 Corvette J52 caliper abutment is fitted to Corvettes with the Z51 brake option. The KORE3 bracket adapts it to the spindle below.

A J52 cross-drilled rotor and caliper is mocked up on a Belltech spindle, to show its size relative to the stock unit.

to the standard G-Body ball joints and has a similar cast-in steering arm. The G-Body tie rod end fastens to the spindle and no changes are required. Belltech and others make a dropped version of this spindle, which is a good idea if the car is to be lowered. Otherwise, an upgraded braking system is as close as the nearest salvage yard.

In stock form, the S-series Blazer uses an 11-inch, one-piece rotor and a dual-piston aluminum caliper. The slightly larger rotor surface and a bigger swept area from the pads provide a noticeable improvement over stock brakes. It still easily fits with most 15-inch and larger wheels. This is a perfect upgrade for someone who wants to keep the original wheels, but wants better brakes.

A longer brake hose is the only non-stock part needed. Use an aftermarket Teflon-lined, braided stainless-steel hose compatible with the stock lines. These aftermarket lines have the standard 10-mm banjo fitting on the caliper end and a standard 10-24 double flare on the chassis end. The stock S10 hoses are not compatible on G-Body cars, even if the length is correct because the chassis end uses a metric bubble-style flare.

Rotors and Calipers from Corvettes

Another upgrade option also uses the 1998–2002 S-series 2WD Blazer spindles but allows the use of C5 or C6 Corvette rotors and calipers. The factory caliper bracket must be replaced with an aftermarket bracket from a company such as KORE3 to allow for this modification. Like the OEM unit, the bracket bolts directly to the spindle, so you can install an upper caliper bracket from any C5 or C6 Corvette.

The rotor size determines the proper abutment. A standard C5, C5 Z06, or standard C6 all use the same abutment. C6s with the Z51 suspension option use a larger, 13.3-inch drilled rotor and a slightly taller abutment (to make up for the larger rotor). The Z51 setup was used on our project car, the *Grocery Getter*, with great success.

Unlike an aftermarket rotor or caliper, these Corvette parts are readily available at dealership parts departments and auto parts stores. Since many Corvette owners upgrade their brakes, these parts are often easy to find in good, used condition, but even new they are surprisingly inexpensive. The C5 base brake system can be purchased for as little as $500, with the only used parts being the spindles and hubs. As with the stock S10 Blazer brakes, a different brake hose is required.

Brake Proportioning Valve

Brake proportioning is often an afterthought for most, but it is an integral part of any brake system. In many cases, the stock proportioning valve (or, more correctly, combination valve) is adequate for stock and mildly upgraded brakes. I generally discard the stock valve on any vehicle, and install an aftermarket proportioning valve when upgrading the brake system.

In some cases it is more for packaging purposes and aesthetics than anything else. However, sometimes it's difficult to find an ideal location for the aftermarket valve. On some vehicles with headers, particularly big-blocks or LS-series engines with large primary tubes, it is located inside the frame rail. Even in cases where it doesn't directly interfere with the headers, it can heat up and boil the brake fluid. For this reason, I locate the valve on the underside of the frame rail, with the adjustment knob facing outward so that the brake bias can be adjusted easily with the wheel turned to the right, in full-lock position.

An adjustable proportioning valve is used to change the bias to the rear brakes, whether due to an upgrade to rear discs or just for fine tuning. These simply consist of a block with an adjustment knob and inlet and outlet ports. They are installed in the rear brake line, and are available from Wilwood and others.

Some provision needs to be made for teeing the stock front brake lines when replacing a typical factory-style valve. This is easily done with a few fittings from the local parts store, but a cleaner way to do the same job is with an adjustable combination valve. Wilwood makes a nice one (PN 260-11179) that allows a much cleaner plumbing job, as well as having a provision for a brake-pressure warning light that can be wired into the factory dash light. They come with fittings, but if you need extras, they are 3/8-24 IF (inverted flare), for 3/16-inch tubing.

A line lock (Hurst calls it "Roll Control") is another effective upgrade for doing burnouts, rather

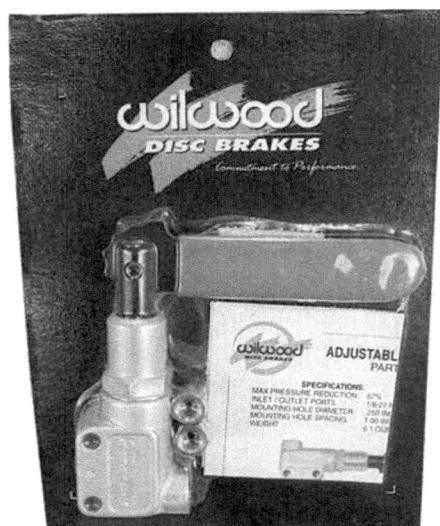

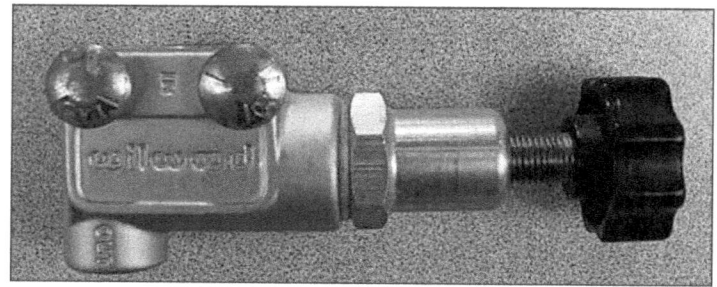

Wilwood offers premium-quality proportioning valves in two styles: knob (right) or lever (left).

I prefer to use Wilwood's adjustable combination valve for G-Body builds when deleting the factory proportioning valve. The factory valve is in the way of many headers (on the inside of the frame rail), and is set to work with the factory brakes. Aftermarket valves allow you to adjust the line pressure to the rear as needed for better braking (since some caliper designs/bore sizes require more pressure than others).

Moroso produces this particular line lock, but many others are produced by various brands.

than the typical "power brake" that destroys the rear brake linings in the process. An electric solenoid is installed in the front brake line from the master cylinder, and it allows the driver to hold pressure on the front brakes, effectively locking them independently of the rear brakes.

Installation is simple, especially if you are replacing brake lines anyway. After finding a suitable place to mount the solenoid, you run a new brake line from the master cylinder port that controls the front brakes. A rear port is used on a G-Body master cylinder, and it has a larger reservoir if using a different master cylinder. Wilwood master cylinders use the front port for the front brakes. A line runs from the master cylinder ports to the inlet of the solenoid, and then another line runs from the outlet to the combination valve.

In most cases, line locks use 1/8-inch NPT fittings, so you need adapter fittings to go to the 3/8-24 IF line fittings. These fittings come with the line lock. Be sure to use Teflon tape or paste on the NPT side of the adapter fittings.

Once the solenoid is plumbed, ground it with the black wire, and route the red wire to the switch and indicator lamp as per the instructions.

Tap into a 12-volt source, and you are ready to bleed the brakes and test the system. Closely watch your fittings for leaks.

Brake Lines

Once you have selected and mounted the braking components, the next issue is the plumbing itself. If you are only making minor changes and do not have a proper double-flaring tool, you can use pre-flared sections of tubing found at any local parts store. If you make more substantial changes, I recommend getting the proper double-flaring tool, a tubing cutter (some prefer a cutoff wheel on a die grinder), a tubing bender, and a small file to dress the cut tubing.

Equipment Options

You can get a double-flaring Craftsman tool from Sears or order one from Summit Racing (PN 4503) or another mail order supplier.

If you choose the tubing cutter, get one with a built-in reamer; if using the cutoff wheel, you need a

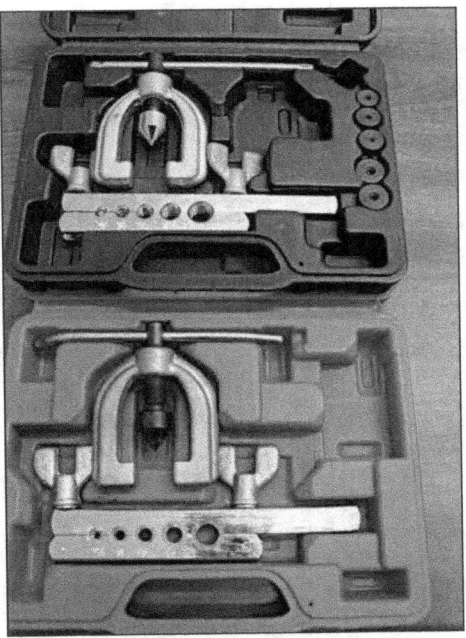

A 45-degree, double-flaring tool is shown at the top; a 37-degree flaring tool is shown at the bottom. It is easy to get the parts mixed up, so be careful if using both on the same project.

A button or shifter handle is often used to activate the line lock. The driver often holds a button on the end of a curled cord while in staging.

TECH TIP: 45-Degree Double Flare

To install a 45-degree double flare, start with the top of the tubing sticking out at the same height as the base of the mandrel. Then slide the desired fitting on the tubing and use a pair of pliers to securely tighten the wing nuts.

Place the tip of the mandrel into the tubing (if you get much resistance, the inside of the tubing needs de-burring), and then use the U-shaped press portion of the flaring tool. Be sure it is straight, or you get an uneven flare.

Tighten the tool so the sharp tip presses a flare into the end of the tubing. Loosen the press enough to remove the mandrel. Then tighten it again to do the second part of the double flare.

Remove the press, loosen the wing nuts, and inspect the flare. If it is uneven, cracked, or looks thin, cut off the flare and re-do it.

TECH TIP: 37-Degree Single Flare

Here, there is no mandrel and only one flare, making this easier than installing a 45-degree flare.

Install the tubing in the appropriate hole in the tool (don't forget the fitting), just as with the 45-degree tool, but instead leave a small portion of the tubing sticking out. There isn't a guide here as with a double flare, so this may take some practice to get it right. Tighten the wing nuts, press in the flare, and you're done.

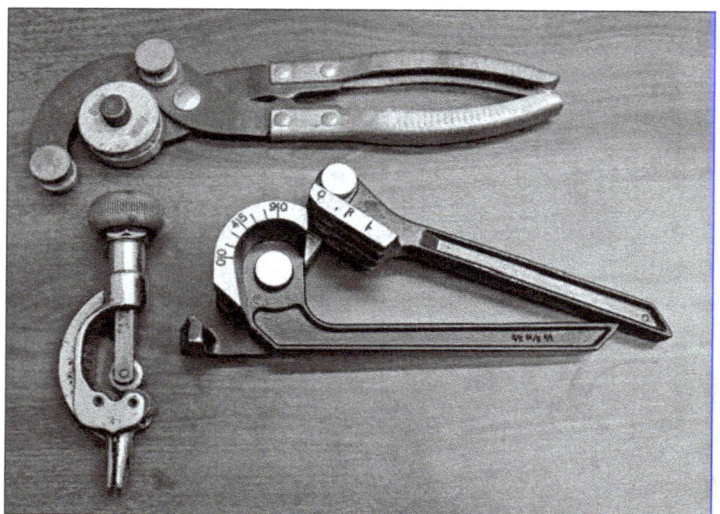

The top unit has an interchangeable wheel for the different side tubing. I prefer to use a cutoff wheel, followed by a reamer, before flaring lines. The tubing cutter work hardens the tubing, making it prone to cracking.

Installing AN Fittings on a Hard Brake Hose

When using AN flares on hard line, two separate pieces form the fitting. The hex portion of the fitting, called the tube nut, is threaded onto the AN thread. To seal the flare to the fitting, you need a separate piece called the tube sleeve. To install them, drop the tube sleeve into the tube nut, with the smaller diameter portion facing downward.

Once it is seated, the smaller portion sticks out the bottom of the tube nut, and you are ready to push it onto the tubing. Install the two pieces onto the tubing before flaring. Tightening the female tube nut onto the male AN fitting seats the larger portion of the tube nut against the back of the flare, and creates a leak-proof seal if the flare is good.

Installing Aftermarket Brakes

The woefully inadequate stock front brakes should be not be used in any high-performance application. These are headed for the scrap heap, along with the stock spindles.

The caliper has been removed. Remove the wheel stud nut and lift the rotor off the hub. These parts will be scrapped.

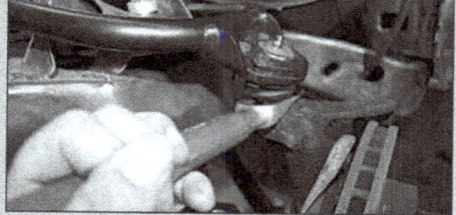

Remove and discard the spindle as an assembly because none of it will be reused. But you should remove the caliper because you may be able to use it for core credit on the new calipers. Simply unbolt the two retaining pins, and remove the line.

MAXIMIZING STOPPING POWER

When installing the new spindle and hub assembly, tighten the castellated nuts, and use new cotter pins for safety.

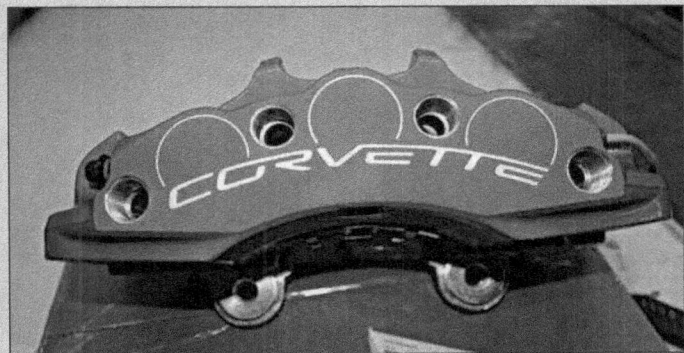

New Z06 calipers are powdercoated red; Grand Sport calipers are powdercoated blue. Otherwise, they are identical.

The C6 Z51 (J52 option) brake rotor, the appropriate abutment, and a C5 caliper are mocked up here. The pad hasn't been installed yet because this is only a mock-up.

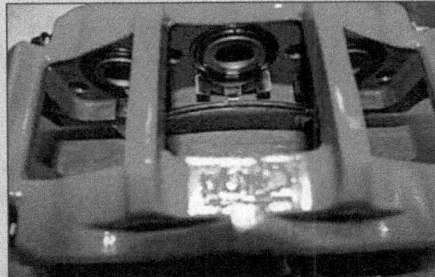

The Z06 caliper has three pistons per side, and each uses a separate pad. Each caliper uses six brake pads rather than two pads as in conventional brake calipers.

The Z06 rotor and caliper provide exceptional brake performance. They are suitable for a variety of high-horsepower G-Body builds.

The Z51 is shown on the left; the Z06 is on the right. The Z51 brakes are more than adequate for the majority of applications.

The $100 bill placed on the C6 Z06 rotor and caliper is for size comparison, not to represent how much they cost or what I spent. These things are very large, and also the best deal around in 14-inch brakes.

separate reamer. A single flare *is not* acceptable for standard 45-degree automotive flares.

Tubing benders are available in numerous designs, and you can never have too many; the main thing to look for is scissor-style handles that are long enough to get the leverage you need when bending larger lines. Summit's version (PN 900156) works well, and isn't too expensive. Its K-Tool (PN 72330) tubing cutter does a nice job, and has the built-in reamer that most hardware store models lack.

Alternatively, a 37-degree single-flaring tool is needed to use AN fittings, such as AN 3, or 3/16 inch.

Tube Bending

Standard automotive brake fittings have a 45-degree cone and are usually made of soft brass or steel. Be sure to purchase flare fittings and not NPT. When installing them, do not use Teflon tape. It is a sealant for pipe thread, not for flare fittings. If you have a leak, look for problems at the flare itself. Also, use flare nut wrenches. Avoid using an adjustable wrench or standard wrenches because these damage the fittings.

Various methods can be used to bend tubing; however you choose to do it, be sure the tubing is not kinked or under tension. I usually make a template out of heavy wire to get the length and shape right, and it acts as a guide for cutting and bending the actual tubing. Although it's possible to form gradual bends in mild-steel brake tubing without kinking, I don't recommend it.

A tube-bending tool is a necessity for tight bends. Several types of benders are available, and no one is clearly the best; each has its advantages and disadvantages, but any of them give you adequate bends.

Brake tubing is usually made of either mild steel, or stainless steel. Inexpensive and premium tools can be used to bend mild steel. Most have a protective coating that prevents rust, at least for a while, in most climates. Stainless tubing has the advantage of increased corrosion resistance, and it can be polished to a high luster, but it is very difficult to cut and flare with hobbyist-quality tools.

Stay away from stainless tubing, unless you are willing to invest in the professional-quality tools needed to properly flare it. Expect to pay at least $300 for a flaring tool that can handle stainless. With cheaper tools, the tubing slips out of the tool before the stainless flares.

Stainless, being much harder than steel, also wreaks havoc on the cones in brass junction blocks. Do not use brass junctions with stainless tubing.

AN -3 tube nuts and tube seals are used on this 3/16-inch brake line. Selecting the proper size of fitting is simple, you just divide the AN size number by 16.

AN tube nuts and tube flares can be used along with the proper union to connect hard lines.

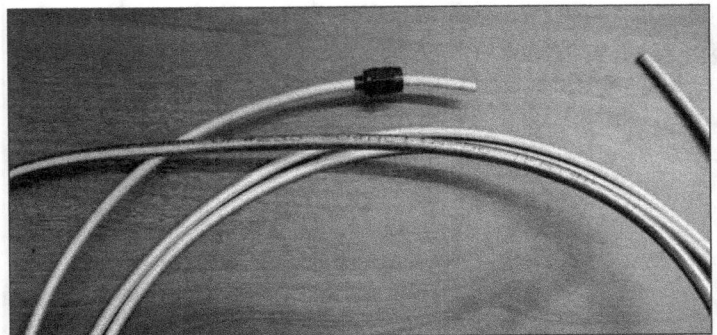

Seamless stainless-steel tubing (top) and mild-steel tubing (bottom) are shown for comparison. Never use copper tubing for any automotive application. The AN tube nut and sleeve are positioned over the mild-steel tubing in preparation for flaring.

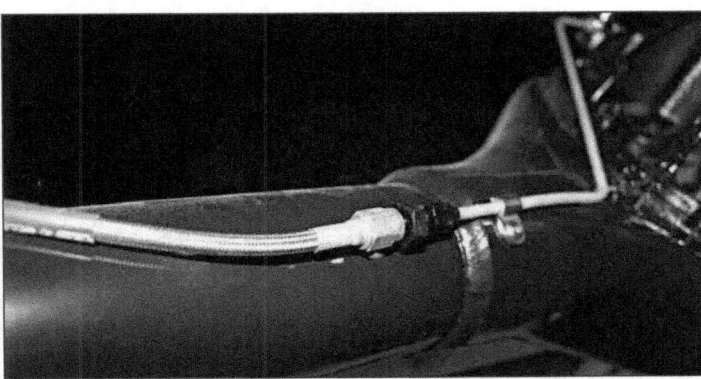

This Teflon-lined brake hose is on a rear disc brake application. It offers a little more abrasion resistance, as well as a firmer pedal, due to less deflection of the hose under pressure.

The 37-degree part is very important; not just any single-flaring tool is sufficient. They are a little harder to find than a 45-degree tool, but readily available through mail-order sources. I typically use double flares at the combination valve and line lock, if applicable, for simplicity (rather than using six adapter fittings), even when using AN elsewhere. Speedway Motors is a great source for anything AN related, and they have a wide assortment of 37-degree flaring tools in all price ranges.

The metric bubble flare is the other type of automotive flare used in brake systems, but this is for specialized applications. In fact, the bubble flare isn't used on stock G-Bodies. Rather than aftermarket braided lines, it has bubble-style cones on the end of the hose that connect to the hard line for Corvette or fourth-generation Camaro calipers.

I have a bubble-flaring tool in my shop, but I recommend using stainless-steel braided Teflon lines from an aftermarket supplier, such as KORE3 or Earl's Performance Plumbing. That way, you have a standard 45-degree cone and you're not adding more complexity to the plumbing than necessary. If using AN fittings, I spec a hose with a -3 female end, and use a -3 union to splice it into a hard line.

Flaring tools vary from manufacturer to manufacturer, but the hand-held versions are basically the same.

To use any of these tools, secure the end of it in a bench vise, and then insert the tubing in the proper size hole. Be sure the tubing is cut cleanly, is straight, and is free of burrs inside. A file and a reamer can take care of any minor issues.

Hose Material

Typically, brake hoses are either stock-type rubber or an aftermarket braided stainless steel over a Teflon lining. Stainless-steel hoses provide superior strength, durability, and line pressure for high-performance applications, while rubber hoses can chafe, crack, and expand. I prefer stainless-steel hoses because they deflect less than rubber under usage, and help increase the amount of fluid pressure to the wheels. They resist chafing, though they should *always* be properly secured.

Several manufacturers offer Teflon hoses. Some sell hoses by length and fitting type rather than application, and they are not DOT approved. These hoses are intended to be used on racing vehicles, which see more demanding conditions than any street-driven car.

So why can't they be used legally on the street? DOT-approved hoses have gone through several DOT tests, including a "whip test." Unapproved hoses are not tested because these tests are prohibitively expensive. Some manufacturers see no need to subject every part number they sell to these tests.

Brake Fluid

While most applications don't need an exotic fluid, a regular flushing of the brake system is needed to eliminate any contaminants and built-up moisture, which ensures longer life for the brake components and lines. I typically use Ford heavy-duty DOT 4 fluid for street cars and Wilwood or Castrol fluid for cars that are going to see extensive track duty. Check with your brake system manufacturer for recommendations based on intended usage.

In any case, stay away from DOT 5 silicone-based fluids. They do have the advantage of not harming automotive paint, so they are great for show cars that aren't driven, but they are more hygroscopic (they absorb moisture) than regular fluids, and they tend to give a spongy pedal feel.

1998–2002 F-Body rear discs use an internal drum emergency brake setup that is easily adapted to other vehicles. These are still available from General Motors for about $100 each.

The 1998–2002 F-Body rear caliper has been installed. While this installation is on a second-generation Camaro, installing the caliper on a G-Body is similar.

Line Bleeding

Once the brake system is installed, bleeding trapped air out of the system is the last step. Nearly every "expert" has a different method, but on a new system, I generally use my Phoenix Systems injector. It can perform typical vacuum-style bleeding (as a MityVac can), but it also does RFI, which essentially forces the new fluid up through the bleeder screw to the master cylinder.

Air naturally wants to rise, so this is a great way to get stubborn air pockets out of the lines. In most cases, use the tried-and-true, two-person method, which involves one pumping the brakes, the other opening and closing the bleeder screw.

Brake Conversions

General Motors saw fit to equip all G-Body cars with drum brakes, and no OEM rear discs bolt on with-

The drum assembly has to be in place before the bearing is installed if you're using a Ford 9-inch or any differential with press-on axle bearings because the backing plate serves as the bearing retainer. In this case, Moser built a bolt-in Ford 9-inch for Chevy-style housing ends. This step isn't needed on C-clip axles (such as the 7.5 or 8.5) that came in the car originally.

out a slight modification to the housing ends and the shock mounts. Once these modifications are made, the brake conversions are simply a bolt-on procedure, other than the required parking brake cable modifications. Many OEM rear discs could be adapted, but here I concentrate on the four most popular swaps.

Something to keep in mind when doing any of these rear disc brake swaps is that while the rotors themselves aren't much larger than the stock drums, the calipers can cause clearance issues on some 15-inch or smaller wheels. The third-generation/early fourth-generation F-Body brakes offer better clearance than the bulkier calipers found on other models. Some clearance grinding can be done if needed, but use discretion.

PBR Single-Piston Calipers

The easiest swap is converting stock G-Body brakes to finned PBR single-piston calipers as used on some late third-generation F-Bodies and all 1993–1997 F-Bodies. The calipers are also found on C4 Corvettes, but the brackets are different to accommodate the Corvette's independent rear suspension.

This 1998–2002 F-Body aftermarket rotor from Summit Racing provides superior braking performance over stock G-Body discs.

Late-Model S10 Trucks

These models have an integral parking brake in the caliper so it's easy to swap (and reducing the cost of new parts) because there is not a separate drum brake assembly inside the rotor as used on others. While caliper brackets must be purchased from a used parts vendor or salvage yard, all other parts can be obtained new from a parts store. Some late-model S10 trucks were equipped with a similar disc brake system; I have never adapted one to a G-Body, but the basics should be very similar.

1998–2002 F-Bodies

The 1998–2002 F-Body brakes are also a very popular swap, with the main differences being a slightly larger 12-inch rotor, a larger caliper with greater pad swept area, and a separate drum brake assembly, which doubles as a caliper bracket. Replacement brake shoes are available from most parts stores; complete drum brake assemblies are available through online sources such as RockAuto, or your local Chevrolet dealer.

C5 and C6 Corvettes

C5 and C6 Corvette rear disc brakes seem to be a logical choice for conversion, especially when using similar brakes in the front. However, the parking brake assemblies can't be used with the stock housing ends. Some use the Corvette parts without a parking brake system, but I don't recommend it. All vehicles with rear disc brakes should have a functional parking brake.

Aftermarket Brake Systems

If you don't want to trust your brakes to salvage yard parts, Hydratech Braking Systems offers complete Hydroboost systems. These are based on new or rebuilt components that are a direct bolt-on for many applications, including the G-Body. While units may seem a bit pricey, these brake kits are actually reasonably priced when you figure in the entire cost of all components. In reality, it's not difficult to spend nearly as much refurbishing and adapting a stock unit.

Even if you purchase all the best brake components for your chosen application, the brakes do not perform as designed unless you use the proper master cylinder. Master cylinder selection for a given brake system is beyond the scope of this book, especially with the myriad braking options available for this application, but here are the basics.

There are two types of OEM dual-reservoir master cylinders: the conventional iron units with a cast-in reservoir and the later "quick take up" style that is constructed of aluminum, with a separate plastic reservoir. Do not use the conventional iron units on the G-Body, unless it has received a B-Body spindle/caliper conversion or some other earlier style of caliper. The aluminum master cylinders are properly calibrated for metric-style calipers, but also work well with most OEM-based

The G-Body "Quick-Takeup" master cylinder has an aluminum body and a plastic reservoir. It rarely has issues and can be used with virtually any brake upgrade other than the B-Body brakes. They require a conventional cast-iron or aftermarket aluminum master cylinder.

Wilwood makes high-quality, aluminum-bodied master cylinders in various bore sizes. Depending on which aftermarket (or retro-fitted factory) brakes you use, an aftermarket master cylinder in a different bore size may be necessary for optimum braking performance.

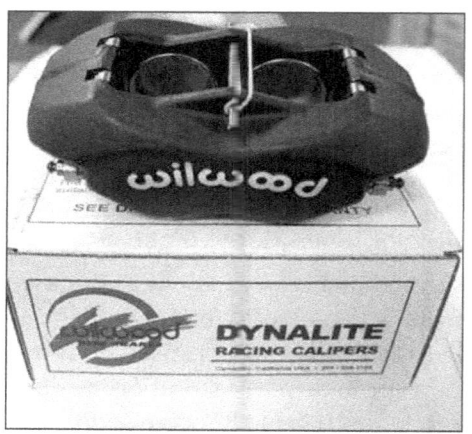

This is a four-piston Wilwood Dyralite caliper, which works very well for a multitude of applications.

Wilwood also has two other front disc brake kits for G-Body high-performance street and track enthusiasts. The 12.88-inch (PN140-12298) and 14.00-inch (PN140-12299) kits feature Forged Narrow Superlite 6R differential bore, radial-mount calipers. Calipers work in conjunction with GT asymmetrical face slot pattern rotors.

Both kits come with aluminum hubs, bearings and seals, hats, radial mount caliper brackets, and all necessary hardware. Optional SRP drilled and slotted rotors and red powdercoated calipers are available in all kit configurations. Wilwood BP-10 high-performance street pads are standard in all three kits. Other brake pad compounds, including those for off-road usage, are available.

These four-piston Wilwoods easily clear the 17-inch Torque Thrust II. This can be a problem with some setups, so always request templates from the manufaacturer.

upgrades, such as C5/C6 calipers, and even rear disc conversions.

As mentioned previously, all G-Body cars came with some form of power assist, but a manual master cylinder from an S-series truck is very easily adapted. In most cases, the brake pedal is attached to an additional hole located below the one used for power brakes. It provides more pedal leverage than the power brake hole, and the pedal mounted in this position delivers a very sensitive brake feel. The master cylinder easily bolts in place of the old booster, but some modification is required for the clevis/pushrod.

Although there are few manufacturers of aftermarket master cylinders, Wilwood aluminum-bodied master cylinders are the most commonly used. Some use these very high quality units for aesthetic purposes or as a complement to a Wilwood brake kit, which requires a different bore size than the stock master. These units dissipate heat much better than older iron units, which is a very good reason to use one. Obviously, extreme heat in the engine compartment from a turbo or supercharger can make any

MAXIMIZING STOPPING POWER

This Baer setup is on Rick Bejerano's Monte Carlo SS. The rear system is very similar to the 1993–1997 F-Body setup. It uses compact, but effective, PBR calipers.

Scott Walkowiak's T-Type GNXray has Baer brakes front and rear, but uses larger monoblock calipers. The monoblocks are more rigid than a typical cast caliper, since they are machined out of a solid block of aluminum. The wheels are from CCW Classics.

underhood plastic become brittle, and the plastic reservoir of a stock master cylinder is no different.

An aluminum master cylinder is a valuable and necessary component on many high-performance cars, and the following story demonstrates why: An acquaintance of mine was driving at high speed on a controlled course when the high-pressure fuel line to his engine came loose. The loose line sprayed fuel at 70 psi all over the engine's hot headers, and flames erupted from under the hood. When he tried to stop, he had no brakes because the heat had already melted the plastic master cylinder reservoir. Fortunately, he used the mechanical emergency brake to slow the car, and he escaped with no injuries.

When the car is rebuilt, undoubtedly it will have an aluminum master cylinder and a fire suppression system as well as a different quick-disconnect fitting. These are wise upgrades for any vehicle that is used in competition, whether the rules require it or not. Aftermarket disc brake kits or components for the G-Body are available from a variety of sources. Wilwood and Baer are the undisputed leaders in performance brakes. KORE3, an established high-performance brake company in the pro touring world, is the best source for OEM-based disc brake upgrades.

I do not recommend other vendors you often see in full-page magazine ads. Many handle Chinese-made, off-brand brakes that are subpar, and the engineering and component quality isn't nearly as good as industry-leading brands. Even those sold under traditionally American product lines, or by large, American mail-order parts dealers, can be very poor quality.

Wilwood Engineering

Wilwood was probably the first commercially successful aftermarket brake company to offer its own calipers. Before Wilwood the only real options in calipers were OEM based. Although some of the OEM calipers of the 1960s and 1970s were very good, they were also very heavy.

Wilwood also has a line of two-piston calipers, designed to replace factory stock calipers with no other changes. They are made of aluminum, and offer weight savings as well as increased stopping power. This caliper is for a B- or F-Body spindle, though a metric version is also available.

Wilwood pioneered the aftermarket aluminum caliper, offering it in multiple-piston calipers to suit virtually any purpose, from drag racing to NASCAR. These fixed calipers delivered much improved braking response over a floating design or typical OEM caliper.

Wilwood offers kits built to order for virtually any application, with varying numbers of pistons and rotor sizes, as well as offering its parts individually for those wanting to build a truly custom braking system. Wilwood also offers its own master cylinders (for both brake and clutch), pedal kits, adjustable proportioning/combination valves, and even brake fluid.

Baer Brakes, Inc.

Baer began to fill a niche in the performance brake market during the 1980s, when its PBR calipers were introduced on the C4 Corvette and 1LE F-Bodies. These calipers were lightweight and relatively compact compared to previous Corvette calipers. Dual-piston calipers were easily adapted to other applications. Baer saw this opportunity and created bolt-on kits that were not only complete but pre-assembled on a modified OEM spindle.

Baer still offers many PBR–based designs, but also offers kits based on other calipers made by Alcon, Brembo, and others. Like Wilwood, Baer has developed its own monoblock caliper for the most demanding applications.

KORE3 Industries

Unlike Baer and Wilwood offerings, the KORE3 systems are based almost entirely on factory brake systems from other vehicles, such as F-Body cars or Corvettes. Unlike most other brake component companies, KORE3 sells all parts in kits as well as individually, including the specific brackets, hardware, and hoses.

So, if you want to bolt on a complete set of 14-inch ZR1 brakes to a 1987 Monte Carlo SS, you can order it as a complete bolt-on kit, or just order the bracket kit and hoses and use the brakes you scored at the last swap meet.

The KORE3 system for the C5/6 Corvette brakes offers some convenient features. Changing the abutments and rotors allows for a switch from standard C5 12.5-inch brakes to C6 Z51 13.3-inch brakes. Therefore, if you are planning a wheel upgrade in the future, you don't have to buy a lot of new parts to get the bigger brakes.

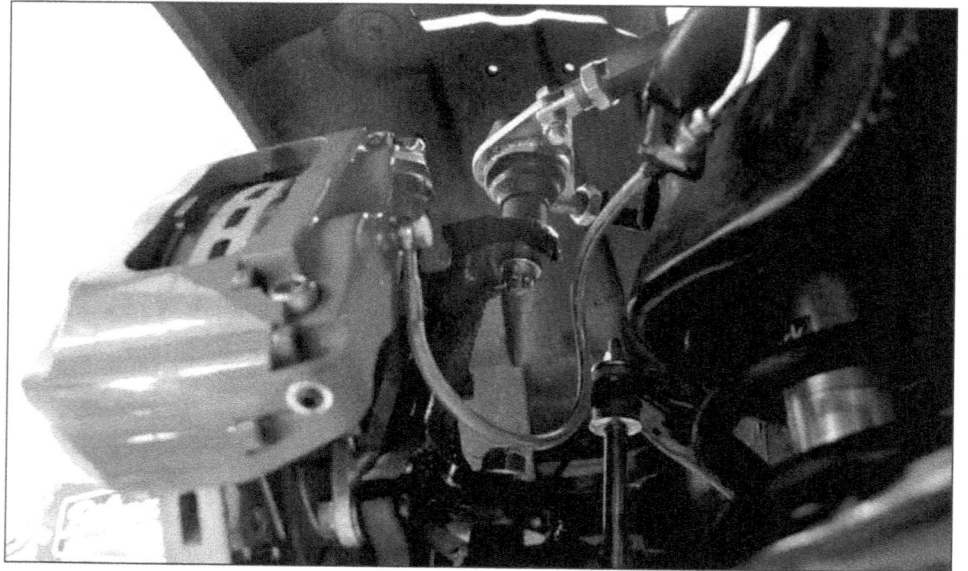

The four-piston Baer brakes on Doug Lutes' Sic Monte. No fairground queen, Lute beats on this car mercilessly at autocrossing events all over the East Coast.

CHAPTER 4

GETTING A GRIP
FRONT SUSPENSION PERFORMANCE

Most car enthusiasts are familiar with the GM G-Body's front suspension because it is similar to every GM vehicle built since 1955. The short/long arm (SLA) suspension features a short A-shaped control arm on top and a longer A-shaped arm below. But this suspension can be substantially upgraded to perform at a much higher level than stock. The arms on 1982-and-newer S-series trucks conveniently interchange with these stamped-steel OEM arms. In fact, the lower arm is also interchangeable with the one found on 1982–1992 Camaros and Firebirds. They are generally rebuildable, but check to make sure the arms have not been damaged in a collision or by excessive corrosion. If so, discard them.

A coil spring is captured in a pocket in the frame, and it seats in the lower control arm.

A conventional hydraulic shock absorber is installed through an opening in the lower control arm. It also bolts to the lower arm and the center of the frame pocket.

The spindle or upright is a "short" design, similar to that used on the 1964–1972 A/F/X-Bodies, but it is not interchangeable with those spindles.

All G-Body models were equipped with a solid front sway-bar; sizing varies.

Overall, this is a very simple suspension design, and it lends itself to easy repair and modification.

Component Options

Suspension components wear out over time and need to be replaced. About any G-Body that hasn't already received a full suspension rebuild needs it, even if it's a low-mileage car. Control arm bushings crack or break. Ball joints wear, and play develops

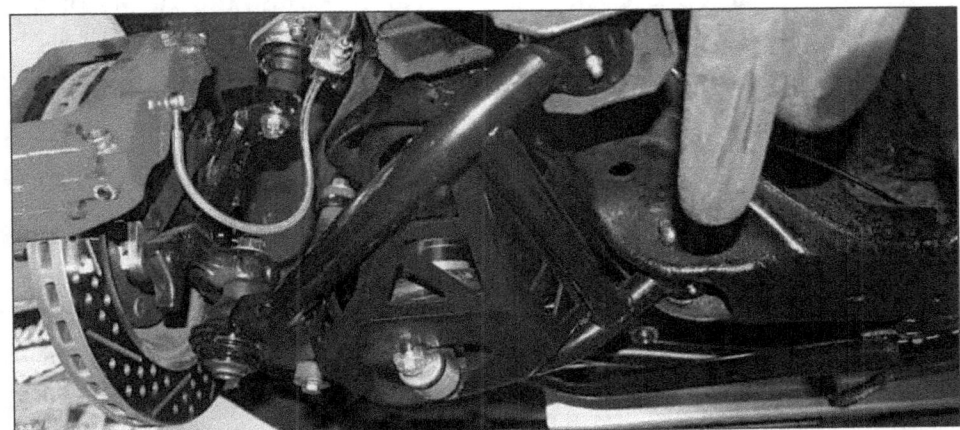

Doug Lutes built his own suspension system. It features SPC upper and lower control arms with Howe screw-in ball joints, Chris Alston ChassisWorks Vari-Shocks, and Moog springs. RideTech coil-overs, however, have replaced the Moog springs. The sway bar is a 36-mm hollow F-Body unit from Spohn Performance, and the end links (note the red bushings) are from Energy Suspension. Lutes has selected some of the best parts from the aftermarket, and created a well-balanced system that suits his purpose (autocrossing) very well. (Photo Courtesy Doug Lutes)

CHAPTER 4

This is the stock front suspension from our project Buick, the GNXcess. It is very similar to that of the earlier A-Body models in both design and construction. The upper and lower A-arms are stamped steel, with rubber bushings and a conventional spring and shock.

Everything is original, right down to the ball joint. All of it will be discarded for this build.

(detected by jacking up the front of the car, and trying to manually move the wheel/tire assembly; if it has play in it, replace the ball joints). Split boots on the ball joints or tie rods occur over time. All are reasons to carefully inspect the suspension and replace parts as needed.

Many enthusiasts rebuild stock control arms, rather than installing aftermarket tubular arms. Some replace stock arms with tubular arms, and wrongly assume the aftermarket versions are stronger and lighter than stock. But, in many cases, the stock arms are actually stronger and lighter than the pricey aftermarket ones. Aftermarket tubular arms have distinct advantages, but always carefully examine them and compare them against the stock arms. (This is covered in "Aftermarket Control Arms" on page 51.

The thing to remember is to use only high-quality parts. If available, use Moog suspension parts because they are made to exacting standards. While these parts are more expensive than store brands, this isn't a place to skimp because the quality of the parts influence the quality of handling, road holding, and ride. Offshore ball joints often fit loosely, and even come apart, so do not use them.

It's matter of personal preference when selecting bushings. OEM-type rubber provides a good ride, low noise, and a reasonable service life. Polyurethane bushings tend to have less deflection and are often chosen for high-performance applications. They resist cracking better than rubber, but can be noisy if improperly lubricated. Also, they can cause binding in some applications, which mostly occurs in the rear suspension due to the already-short length of the upper control arms and their relative lack of leverage. Bushings made of DuPont Delrin (often paired with aluminum as in Global West's Del-a-lum bushings) have the advantage of virtually zero deflection, but they are costly, and also tend to squeak if improperly lubricated.

Solid aluminum bushings, steel bushings, and heim-joint ends can replace conventional bushings, but these are installed on aftermarket arms; rarely, if ever, are these parts installed on stock arms. They tend to be noisy and wear more quickly than other choices. Keep in mind, heim ends need to be inspected regularly. I include them only because some aftermarket arms may offer them.

Stock Control Arms

Rebuilding stock control arms is pretty simple and straightforward, but you need air tools and a hydraulic press to complete the work. If you do not have these tools, leave this job to a professional. Most home mechanics can remove the arms, but always exercise safety when doing the job. The lower control arm contains the coil spring, and the compressed spring has enough energy to severely injure or kill if it slips out of the compressor.

Performing this work is much easier with a car lift. If you don't have one, use a floor jack to raise the front of the vehicle. Place jack stands securely under the frame and around the firewall area. Be sure to get the car high enough so that the lower arm has room to swing down without hitting the floor.

Here are the general steps I follow:

1 Once the car is in position, remove the front wheels and then unbolt the sway-bar end links from the lower control arms. The links are often badly corroded and have a tendency to break, so use lots of penetrating oil and be prepared to replace them. 2. Cut them off with a plasma cutter or reciprocating saw, if needed.

2 Use an air ratchet to remove the upper retaining nut from the shock absorber, and then remove the two bolts at the bottom of the lower control arm. The shock can be lifted out of the suspension.

3 Next, use pliers and hammer to tap and pull the cotter pins out from the upper and lower ball joints, as well as from the outer tie rod end. Note the size and discard them. *Never* reuse cotter pins because if a reused cotter pin fails, a vital suspension fastener could come loose and suspension failure could occur.

4 If brake work is being done at the same time, disconnect the flex line from the caliper. I always do this at the caliper because there is less chance of stripping one of the soft fittings; otherwise, just unbolt the caliper from the spindle and use a piece of wire or a bungee cord to hang it from the chassis or suspension, which minimizes stress on the line.

5 Once the caliper is out of the way, loosen the castellated nut on the outer tie rod end. It can be removed, but be sure to choose the right method for the tool you have available.

6 The easiest way is to smack the loosened nut with a small, 3-pound sledge, but it must be left in place. Chances of damaging the threads are minimized and this force is usually enough to break it loose. If this method does not work, use a "pickle fork," or ball joint/tie rod separator, which is similar to a large chisel with a forked end. Strike the end with a hammer to force apart the tapered tie rod and steering arm. This is a very versatile tool, but care has to be taken when using it on good parts because it has a tendency to rip the rubber boots found on tie rods and ball joints.

7 Use a tie rod puller to press the stud out of the steering arm. This tool is best for use on "finished" cars, since it doesn't damage the tie rod boots as a pickle fork does.

8 With the tie rod out of the way, remove the coil spring and control arms. Some use a spring compressor to remove the control arm. I recommend using a floor jack for the spring removal procedure. Place it under the lower control arm, positioned directly under the lower ball joint, and raise the lower arm just high enough to take the tension off the ball joints. If you are not going to reuse the spring, use a plasma cutter, cutoff wheel, or reciprocating saw to cut it out and remove it. Use care and do not position yourself in the spring's path, should it somehow slip loose.

The upper ball joint is riveted in, which indicates it has never been replaced (aftermarket ball joints bolt in). The steering linkage has been disconnected from the steering box and spindles, and the front sway bar has been removed in preparation for control arm removal.

The end link installs in the lower hole of the A-arm. An end link is basically a long bolt with bushings, a spacer, washers, and a locking nut. They are very strong and rarely, if ever, break. When removing them, don't hesitate to cut them off if necessary, especially if the threads are rusty. These bushings are in decent shape with no visible cracks, and therefore are unlikely to be the original parts.

This Belltech dropped spindle for a 1998–2002 2WD S-series Blazer allows the use of larger brake rotors. The upper ball joint nut has been loosened, and a pickle fork has been used to break it free from the spindle. At this point, the control arms securely contain the spring; the shock has been removed. Next, place a jack under the control arm and put the spring under a load. If you don't have much experience with coil springs, use a spring compressor or chain to limit the travel of the spring when removing it. Exercise extreme caution, you do not want the spring to slip.

9 Remove the castellated nut from the upper ball joint. Use the pickle fork and hammer to separate it from the spindle.

Disconnect the upper ball joint from the spindle. This ball joint separator is essentially a large wedge; hit it with a 3-pound sledge to break the interference fit of the ball joint and spindle. The aftermarket A-arm is a StrongArm from RideTech.

10 Once the ball joint has broken loose, remove the two nuts to unbolt the upper control arm shaft from the frame, and remove the arm.

11 Remove the lower ball joint in the same manner, but leave it in place temporarily, so it shields you from the coil spring.

12 Once the lower ball joint is loose, *slowly* lower the floor jack and release the tension on the spring. Often, the spring falls out onto the floor, but in some cases, a little residual tension makes it hang up on the spring seat. If this happens, use a long pry bar and carefully unseat it, but be sure to stay out of the spring's path.

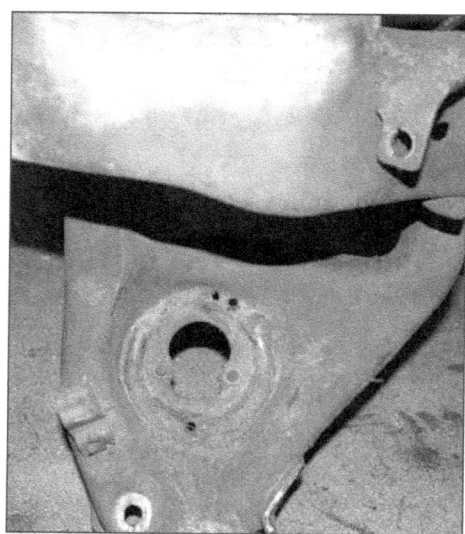

Exercise extreme caution and use the correct tools when removing the spring. With the upper arm disconnected, lower the bottom arm to remove the spring. Even when the spring is compressed, don't do this by hand. Use a long pry bar and keep your body as far away from the spring as possible. The energy contained by a spring can badly injure or even kill you. With the spring out, remove the bolts securing the control arm to the frame, and pull it free.

13 Once the spring is out of the way, remove the two lower control arms and set them aside.

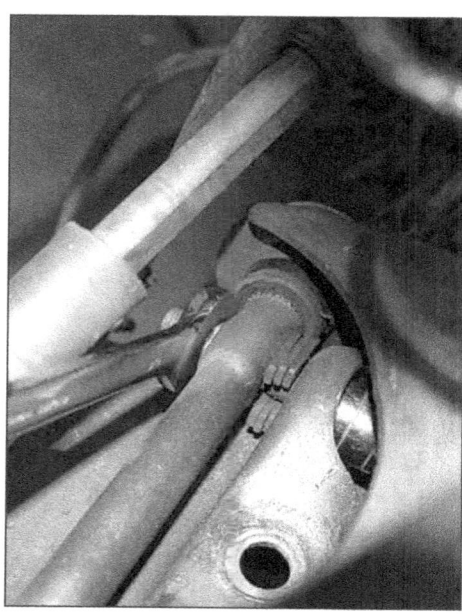

The spindle end is already disconnected, so all that remains is to remove the two nuts at the upper control arm shaft. The upper arm then comes off, and this side is finished.

Box in the stock lower arm with steel plate, and then add a new ball joint and aftermarket bushings for more strength. Adding steel plate to fully box the arm does add some weight but the increased torsional rigidity and improved handling are definitely worth the trade-off. (Photo Courtesy Ben Meissner)

For the upper control arm, you need an air or electric grinder or a drill to remove the factory rivets on the upper ball joint. I follow these general steps:

1 Use a hydraulic press or air chisel to remove the control arm bushings. Grinding the ball joint rivets is usually faster; once the heads are ground flush, use a hammer and punch to drive them out. Replacement ball joints are supplied with bolts, nuts, and lock washers for reassembly. If the upper ball joints are bolted in already, they have been replaced before. Be sure to use thread locker on the new bolts because you don't want to risk these ever coming loose on the road.

2 To remove the upper control arm bushings, first remove the cross shaft nuts and washers. This can be done with hand tools. If the control arm bushings are mostly intact, they can be difficult to remove. A good trick is to use a drill to remove most of the rubber bushing material, and a handheld propane torch to take care of any residual rubber.

3 The bushing's steel sleeve has to be pressed off or chiseled out. If it must be chiseled off, air chisels work best, but it can be done by hand in a pinch. Carefully support the control arm when using a press because it is very easy to bend it.

4 Once the first bushing shell is pressed out, remove the cross shaft. With the cross shaft out, it's easier to remove the other bushing.

5 The lower control arm bushings are removed in the same manner (though there isn't a cross shaft to deal with; you simply remove the bushings and bushing shells).

6 Use a ball joint press to press off the lower ball joints. You can rent one from most auto parts stores. The press works similar to a big C-clamp; it has several round tubing adapters to press the old joint out and the new one in. If needed, do this work when the control arm is installed on the car. You can also use a hydraulic press and some large tubing.

7 Installing the new bushings on upper and lower arms is pretty straightforward. The OEM-style rubber and aftermarket urethane bushings come with new shells that are a *very* tight fit, so they need to be pressed in. Carefully support the control arm so as not to bend it. If necessary, use an abrasive wheel to grind away the ID of the bushing area on the control arm to make the installation easier, but go slowly and be careful. If you remove too much material, you have to replace the control arm.

8 Apply lubricant to polyurethane bushings (and other bushing types that require it) before installation to avoid squeaks.

Aftermarket Control Arms

Stock control arms have their place, but aftermarket arms can have several advantages. As mentioned previously, weight and strength aren't necessarily among them. In many cases, they are heavier than stock, and some arms are weaker than stamped arms. Drag-race arms are weaker, but there is rationale for using them. In most cases, the upper A-arm features improved geometry, but the geometry for the lower arm typically remains the same.

The length is often changed to compensate for a taller spindle. The B/F-Body spindle or newer

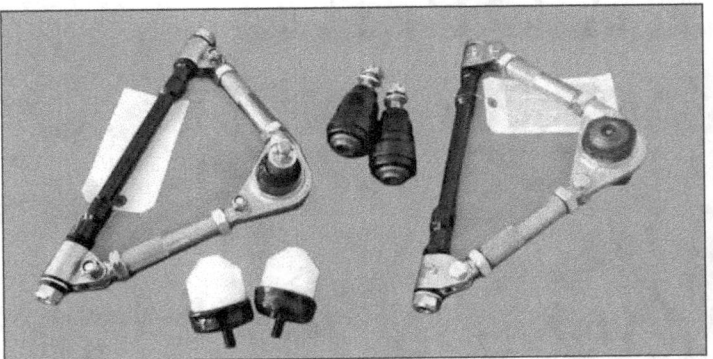

Ben Meissner installed these SPC upper arms. Incredibly, these use a greasable steel bushing. As long as they are kept lubricated, they last forever. (Photo Courtesy Ben Meissner)

Here, an SPC arm is installed on Doug Lutes' Sic Monte Monte Carlo SS. (Photo Courtesy Doug Lutes)

aftermarket spindles for the A/F/X are often taller. Sometimes, though, the geometry isn't changed at all. In this case, the benefit is the arm's bushing material or a dimensionally different/stronger ball joint.

Keep these factors in mind when choosing new control arms. If you need more information, consult an expert such as Mark Savitske of Savitske Classic & Custom. His book *How to Make Your Muscle Car Handle* does a great job of explaining suspensions, and how to get the best results for a given purpose.

Aftermarket Control Arm Installation

I chose a set of upper and lower StrongArm control arms from RideTech Suspension for the *GNXcess*, which is a 1983 Buick Regal T-Type. RideTech's initial efforts were focused on some very innovative air

The GNXcess isn't fitted with stock suspension components. Pictured here is the complete suspension package from RideTech. The new StrongArm lower control arms are compatible with Shockwaves (a hybrid shock) or with coil-over springs. The front and rear Musclebar sway bars are also pictured.

With the GNXcess disassembled, we did a quick mock up to verify the fit of all components. While it's not essential, it's helpful to see how all the parts come together. Upper and lower StrongArms, Belltech Blazer 2WD spindles and hubs, and coil-over shocks have been installed.

In addition, a 525-hp LS3 crate engine from Chevrolet Performance has been installed. But this engine isn't a permanent addition. It is destined for a 1965 Cutlass, the Lonestar. We will install a 408-ci LS engine with twin turbos. Custom-built stainless steel turbo headers by Bulldawg Musclecars will replace the BRP/Musclerod 1⅞ headers.

The RideTech coil-over shocks mount up just like stock versions, but do require a slight enlargement of the hole at the top of the spring pocket for the shaft to pass through.

The Musclebar is a direct bolt-on, but two holes do have to be drilled and tapped for its mounting brackets.

These are similar to late-model Corvette end links (pictured on the A-Body chassis on page 55).

Notice the clearance between the upper control arm and the header; the arm can actually be serviced without removing the header first.

Header clearance is also excellent on the driver's side.

The combination of LS engine, RideTech suspension, and Z06 Corvette brakes will transform the handling and performance of this G-Body.

The RideTech suspension and Belltech 2-inch dropped spindles provide an attractive and aggressive stance for this 1983 T-Type. It may be a little low for a street car, but it works great on the autocross course and at my prime destination for it, the Bonneville Salt Flats.

suspension products that are available for the G-Body. RideTech remains the industry leader in air suspension, but has branched out to offer complete suspension packages for a variety of vehicles.

The StrongArms were originally designed to accept RideTech's Shock-Wave (a combination of an air bag and an adjustable aluminum shock absorber). RideTech's billet adjustable coil-over shock with Hyperco springs are for those who plan to autocross as well as road, drag, and land speed race, as well as regularly drive on the street. Drastic changes that can be made with a conventional spring, or even a ShockWave, are needed.

Made from heavy-wall, mild-steel tuning, StrongArms feature an attachment point on the lower arm for coil-over shocks, and therefore they're not compatible with a conventional coil spring. The improved geometry of the upper arm offers better handling and is set up for a "short" spindle. (Note that my spindle isn't the original G-Body spindle with its cast-in caliper bracket.)

I installed an aftermarket Belltech 1998–2004 2WD S-series Blazer spindle with a 2-inch drop, which is the same height as a stock spindle, because the StrongArms can't be used with tall spindles. The arms come fully assembled, except for the upper ball joints (included), which must be bolted in. The arms are fitted with polyurethane bushings and they feature a satin-black, powdercoating finish. Installation is the same as with stock arms, and the stock hardware is reused.

CHAPTER 4

This RideTech Strong-Arm lower control arm doesn't have a provision for a standard coil spring because it's set up for a coil-over shock or ShockWave airbag/spring. These arms are extremely beefy and come with the ball joints and bushings pressed in. The tab on the left is for the sway bar end-link mounting.

RideTech has air systems that provide exceptional handling comparable to the best conventional systems available from the aftermarket, and the ShockWave is a big part of that. The high-quality adjustable shock within an air spring also solves a lot of suspension and body panel fitment issues, making air suspension suitable for G-Body and many other applications.

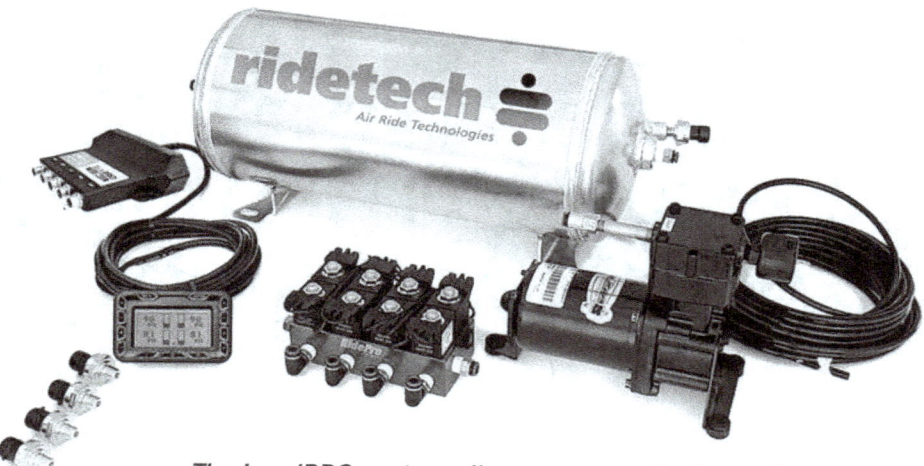

The LevelPRO system allows easy monitoring and pressure changes right from a dash- or console-mounted touchscreen.

RideTech's RidePRO system features this compact air compressor, tank, and valving assembly. The unit easily fits under the package tray.

Sway Bars

Selecting the correct sway bars (technically, anti-sway or anti-roll bars) reduce unwanted body roll and suspension action to yield major improvements in handling. Essentially, the sway bar acts similar to a torsion bar, connecting the frame to the suspension on each side of the car. Therefore, the sway bar needs to be properly matched to the spring rate of the car and its application. If the sway bar has a larger diameter, less spring rate is needed; if the sway bar has a smaller diameter, more spring rate is needed.

Often, sway bar selection is critical and somewhat similar to camshaft selection, so do some research and choose wisely. Bigger is not necessarily better.

Stock Bars

Similar to the front bar on many other GM vehicles, the G-Body's front sway bar is mounted to the frame rails with rubber-insulated brackets and rubber-bushed end links. The end links are basically long bolts with a steel spacer plus two steel washers, two rubber bushings, and a nut. In the rear, the sway bar bolts directly to the inner portion of the lower control arm and not directly to the frame. This design limits the effectiveness of the bar because it has less leverage than a sway bar that attaches directly to the rear end housing. Rear sway bars on the third- and fourth-generation F-Body cars link from the rear axle to the frame, and this provides more effective suspension control and reduced body roll. This design is a carryover from the A-Body.

Fortunately, General Motors offers many sway bar sizes for the G-Body

GETTING A GRIP

and the front bar of third-generation F-Bodies. You may have to try several types and diameters to find the correct size. This enables you to purchase different sizes for very little money from salvage yards, and experiment to see what works best with your particular combination. If you have friends who drag race their G-Bodies, you may even be able to get the sway bars for free because they remove them and store them for later use.

Aftermarket Sway Bars

Most aftermarket sway bars are similar in design to the stock pieces, but often in larger diameters. Oddly enough, most manufacturers only sell one bar for a given chassis. If your car is equipped with manufacturer springs, the bar is matched to the spring rate of those springs. Otherwise, it may not give the desired results.

Larger, polyurethane bushings and matching frame brackets, as well as higher quality end links typically come with aftermarket front bars. In the case of some rear bars, the end link can be installed in several different locations, so you can tune the bar to your particular suspension setup and application. Most suspension tuners recommend starting with this type of bar in its softest setting, then adjusting to a stiffer setting.

To get the best performance ensure that the front sway bar ends are as close to parallel to the ground as possible. If the car is lowered, the end link is often too long, causing bind. In some cases, an end link can be cut down, but a better solution is to buy a replacement end link in the appropriate length. Energy Suspension is a good source for these.

For many years, circle track racers and serious road racers commonly used a fabricated straight/splined bar with correspondingly splined ends. These bars are typically mounted to the frame with a bushing and strap similar to other sway bars. In addition, these are often larger and stronger than stock and attached to the control arm with a heim joint end link. As with any bar, the rate of the bar must be properly matched to the springs, but these are adjustable while the conventional ones are not, so you can change the position of the end link to stiffen or soften the bar. For many years, this style of bar also appeared on high-end, pro-touring cars, yet most are one-offs, rather than a production design.

RideTech offers what appears, at first, to be a splined sway bar for

This 1971 Cutlass S carries a front suspension that's nearly identical to the one going into the GNXcess. Upper and lower RideTech SrongArms and coil-over shocks have been installed. The stout Musclebar front sway bar system has been bolted in and features late-model Corvette-style end links.

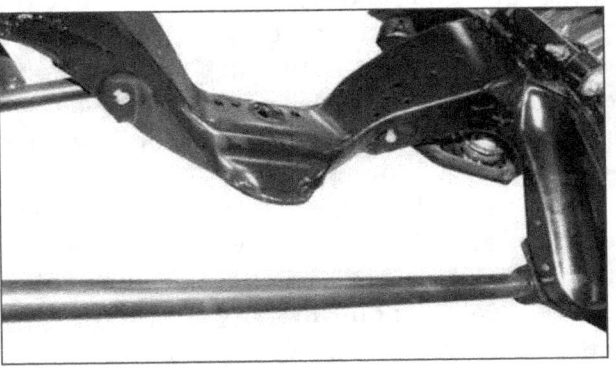

The Musclebar installs in the stock location, but the larger bar size combined with larger tubing requires some re-drilling and tapping of the frame for mounting. You use the bushing brackets as a template to perform this quick and easy procedure.

the G-Body called the Musclebar. It is designed to be used on lowered vehicles, and to maximize tire clearance. Unlike the NASCAR-style bars, RideTech uses an unsplined bar with bolt-on ends that are machined to fit the specific application. Instead of a heim joint end link, RideTech uses Posi-Link end links, which are similar to those found on newer vehicles such as the C5 and C6 Corvette. These give the advantage of immediate reaction, rather than the less immediate reaction of a typical bushing-style end link.

Installation of the front and rear RideTech Musclebars is very straightforward. Here are the general steps:

1 Remove the stock sway bar, sway bar end links, and sway bar mounts and bushings. None of this will be reused, so these can be discarded.

2 Push the sway bar bushings and brackets onto the new bar, and install the supplied hardware. Because of the new sway bar's much larger size, the OEM sway bar mounting holes in the frame are not far enough apart. This is no problem because you can easily drill and tap the two needed holes.

3 Once the holes are finished, bolt the bushing brackets to the frame, and bolt the machined ends onto the bar. Use thread-locking compound on all of these bolts.

4 Bolt the Posi-Links to the machined portion of the sway bar and then to the lower control arm. You may need to put a jack under the lower control arm and raise it to get the link in the proper position.

Sway Bar Removal

Since many G-Body owners are more interested in going fast in a straight line, the number-one front sway bar modification for straightline or drag cars is sway bar removal. Front sway bar removal reduces weight and allows the front end travel to increase on launch, which transfers weight to the rear tires and increases traction in most cases. The extra weight transfer may or may not be desirable, but obviously if you're trying to increase traction the extra weight transfer is desirable. Weight transfer is determined by spring rates shocks, car weight, power output, chassis stiffness, and several other factors.

As with any type of tuning, lots of experimentation is required. A bar smaller than stock may work best on one car, while no bar may be fine on another. For a dual-purpose car, disconnecting the end links from the bar gives much of the same result, and the bar is easily reconnected for the drive home.

Shocks

Shocks are one of the most important components of any suspension, but many people do not take the time to select the proper shocks for the car and application. Shocks have a greater effect on ride and performance than most of the more glamorous suspension parts. A full discussion of shock theory and application is outside the scope of this book, so to simplify things the following are a few specific recommendations.

If you're doing a true restoration, choose stock AC Delco shocks, which can be repainted to look stock. Most people just buy whatever shocks the local parts store has on the shelf, and the average driver never knows the difference. Those types of shocks meet most people's number-one requirement, which is low cost. But by choosing them, owners are leaving a lot of performance on the table.

Budgetary compromises do have to be made, but shocks are one area where I refuse to compromise. There are plenty of high-quality examples. For the average street application (and track applications in some cases), I recommend Bilstein B6 HD shocks (PN 24-009492). This is a heavy-duty, monotube, gas-charged shock.

That is not to say that Koni, QA1, Penske, VariShock, and others don't make a fine product, because they do. But, keeping budget in mind, Bilstein is typically the least-expensive, high-quality shock on the market. If additional features, such as adjustability, custom valving, billet aluminum shock tubes are desired, then the others I mentioned may be a better choice.

Coil-Overs

Coil-over shocks have become an easy bolt-on conversion for G-Bodies. The main advantage of this type of setup is that it allows a far larger choice of spring rates than is typically available, and this lends itself to near infinite tuneability. Does the average enthusiast need this? Doubtful. However, if you plan to autocross or road race, coil-overs are a virtual necessity. Although many companies offer coil-over shocks, RideTech shocks are engineered for the proper mounting brackets to make them a bolt-on installation in any G-Body. RideTech coil-overs are available in single-adjustable, double-adjustable, or non-adjustable versions, and come with Hyperco coil springs. A

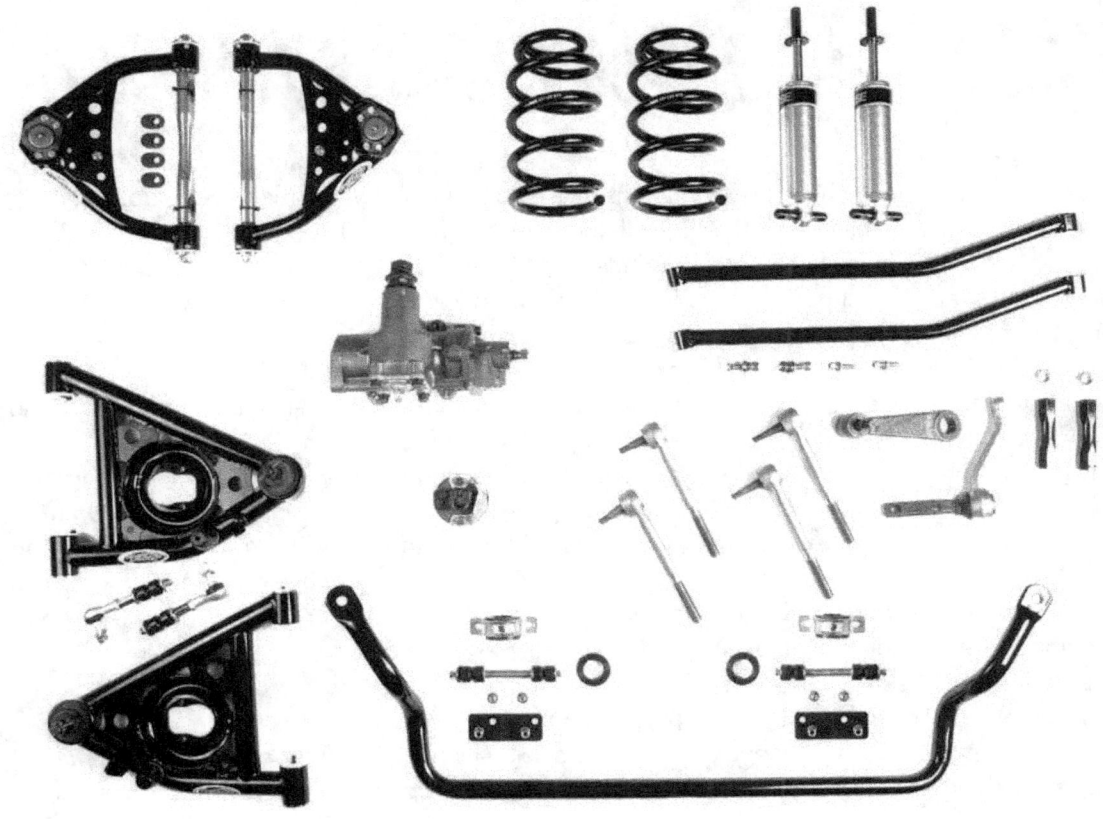

Detroit Speed & Engineering offers a full line of G-Body suspension and chassis products. The DSE suspension system comes with either conventional springs and shocks or coil-overs. The coil-spring version is installed on the GNXray T-Type and provides exceptional performance.

variety of spring rates are available, and RideTech's technical support department can help you decide which is best, based on your application. This is a true bolt-on system, and it comes with everything needed for the conversion.

Springs

G-Bodies use coil springs front and rear, and in most cases, you still find the original springs. Spring rates were typically very soft, at least by modern standards. Most of these cars were not built for performance, and even with performance models, the suspension was still very lacking. This is due mostly to the lack of rigidity, and very compliant bushings and shocks.

The chassis in G-Bodies is very weak and only slightly better than that in the A-Bodies that preceded it, so a higher spring rate (or stiffer spring) is wasted on that particular chassis. This is still true and is something to keep in mind when modifying your G-Body's suspension. An expensive aftermarket suspension doesn't make much sense if the chassis isn't going to be adequately stiffened, so box the frame and add frame bracing before you make any major suspension upgrades.

That said, most G-Bodies have the original springs, and they need replacement. In most cases, they have sagged. Many were improperly lowered through heating the coils or excessive cutting. Or, even worse, G-Bodies were raised to fit a ridiculously large set of wheels. In any case, replacement is simple, and replacing them while rebuilding the front suspension saves time and money.

For a stock application, you can buy springs from any local auto parts store, or an online vendor, such as Rock Auto. Keep in mind, though, that you have little or no choice of spring rate. These are mostly generic springs designed to fit any application. While they do fit, the rate, ride height, and other aspects typically are not ideal, due to production compromises (soft ride) and the wide range of factory equipment. If they work well for you, consider yourself lucky.

A better choice is to order springs from the company that likely made your originals, Eaton Detroit Spring. You can provide the exact specifications of your vehicle (engine size, transmission type, options, etc.), and they can build a new set of springs using the factory blueprints. They cost a little more, but if you want a stock look this is the way to go.

Lowering the G-Body

Most enthusiasts, regardless of how they use the car, prefer a lower stance, especially if the car is equipped with larger aftermarket wheels and tires. Nothing ruins the look of an otherwise nice car more than a horrible stance. Reducing the gap between the wheel opening and the tire does wonders to improve appearance, but this is by no means the sole reason to consider lowering it. A lower center of gravity improves handling, and the right combination of springs and sway bars can drastically reduce body roll while still providing a good ride.

Lowering Springs

Installing a set of lowering springs at the correct spring rate for the vehicle is the easiest way to lower a car. Most aftermarket suspension companies, including Hotchkis, Eibach, ChassisWorks, and Global West, offer these springs. Most suspension companies list the amount of drop that can be expected, but this figure is rarely accurate for a particular vehicle because of the aforementioned variables. You may not see any change at all.

For instance, if the stock springs have sagged significantly, or the new engine is significantly lighter than the original (say an LS3 in place of an all-iron 305), it may retain the stock ride height. Or, the opposite may occur, and the front end is now higher. If this happens, verify that the springs are properly seated in the frame as well as on the lower control arm.

If the springs are improperly seated on one or both sides, they are significantly higher than they should be. If the springs are properly seated and the car is still too high, drive the car for a couple of days before doing anything more. The springs "set in" and that reduces the ride height by a small amount, which may be enough.

If the ride height is still too high, you can cut the springs, but there are a few things to remember when doing this. For one, only cut a spring with a tool that doesn't generate much heat. Use a 4½-inch cutting disc on an angle grinder because it won't affect the temper as does a torch or plasma cutter. Also, cut the spring in no more than 1/2-coil increments.

You may have to remove and reinstall the spring several times before getting the height where you want it, but this is preferable to cutting too much and having to replace the springs. Also, do not ever remove more than a full coil because the spring can be severely weakened and even fail in extreme cases.

Dropped Spindles

If the car is still higher than desired after cutting 1/2 coil from the spring, consider installing a dropped spindle. The typical dropped spindle has a 2-inch drop. The actual spindle portion is raised, which effectively lowers the ride height of the car.

Dropped spindles are readily available for G-Body cars. If using stock spindles, you can select early S10 spindles (Belltech PN 2100) to achieve a 2-inch drop, and these spindles may be advertised as or share the same part number as S10 spindles. If using a 1998–2004 S10 Blazer 2WD spindle, Belltech makes a dropped spindle version for the G-Body (PN 2102).

This spindle is sold under the Summit Racing brand, but is likely made by Belltech.

New OEM-type spindles are available from Speedway Motors and Summit Racing (shown). They are a good option if new or reconditioned ones are not available. These are also an excellent option if your stock spindles are in poor shape and you don't want a drop or plan to do a brake conversion.

GETTING A GRIP

So, how do you determine the best spring for your particular application? Unfortunately, there are so many variables (even factory cars came with hundreds of spring options), and everything that affects the weight of the car ultimately affects which spring you need.

To get in the ball park for a typical street or pro-touring application, start with Hotchkis springs (PN 1902). If using a lighter engine (such as an all-aluminum LS3), you probably need to cut them to get the ride height correct. This requires trial and error. Before cutting anything, though, drive the car and give the springs a chance to settle. It's easy to pull the springs and cut them, but you can't put it back, so be sure of what you are doing. I don't recommend cutting more than half the coil from these springs.

Competition Springs

Special application springs or custom lowering springs with a specific rate are available. If your G-Body is used for autocrossing, road racing, even circle track racing, you may have a specific spring rate target for the particular application. All of these applications have different requirements than the typical, off-the-shelf "lowering" spring for high-performance street use.

The easiest and least expensive way to go is with a race-oriented spring company, such as Afco Racing. While Afco is primarily known as a circle track supplier, the company offers springs in nearly unlimited rates. Selecting the appropriate spring rate is beyond the scope of this book, but your spring manufacturer can often lead you in the right direction. Mark Savitske's *How to Make Your Muscle Car Handle* provides extensive instruction for selecting spring rates for particular applications.

Drag racing applications have their own spring requirements, but many successful racers use lighter springs found in V-6 applications. In many cases, these may be the springs their car originally came with because many G-Bodies were V-6 equipped and later swapped to V-8s. Some manufacturers, such as Moroso, offer a "trick" front spring, designed for increased weight transfer. This type of spring isn't recommended for a vehicle that sees street use, however, due to the rapid weight transfer (in either direction, accelerating as well as braking).

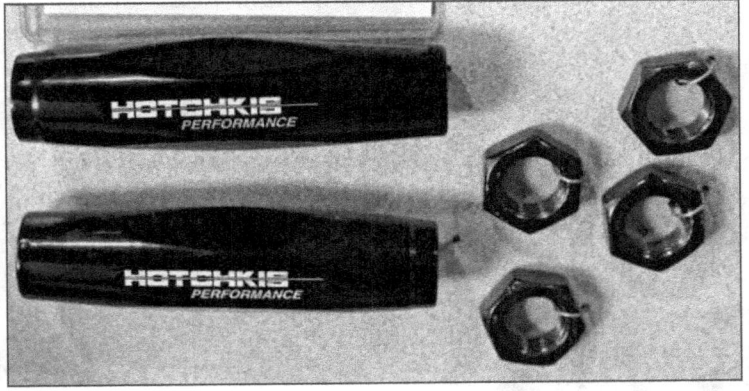

The stock tie-rod sleeves are very weak, and are hard to adjust quickly; these units from Hotchkis solve both problems.

Since this car is street-driven and does autocrosses, the Hotchkis tie-rod sleeves make things a little easier when changing alignment settings. (Photo Courtesy Doug Lutes)

This front sway bar uses Energy Suspension bushings. They are available in a variety of sizes and colors.

CHAPTER 5

HANDLING ON RAILS
STEERING SYSTEM UPGRADES

General Motors equipped the G-Body with a solid steering system, but this system did not provide very good steering performance. And it's certainly not ideal for high-performance applications. In this chapter, I review popular modifications for increased steering performance and reliability.

Jones Racing Products makes this slick power steering pump setup, which uses a custom bracket that attaches to any cylinder head through accessory bolt holes. It uses a GM Type II pump with an internal reservoir. It is plumbed with AN -6 teflon line. (Photo Courtesy Doug Lutes).

Steering Box

All G-Bodies came equipped with a Saginaw-designed recirculating ball-type, integral power steering box, which used an engine-driven hydraulic pump. In fact, the G-Body never had a manual steering option.

The Saginaw 605 power steering box was installed on many of the early G-Bodies. The 605 box is easily identified because it has a distinctive round cover held on by a large snap ring, rather than the typical GM rectangular, four-bolt cover. Most, but not all, have standard flare-type hose fittings, rather than the O-ring styles used on all later boxes. This very light duty box had a horrible ratio (around 17:1), and it was likely discontinued for those reasons. It was mostly a weight-saving measure, but General Motors saw the error of its ways and only used the box for a few years. The old-standby 800 series box replaced it.

Unfortunately, the aftermarket does not offer support for the 1960s steering box or parts for it, and nothing can be done to improve it. If you have a 605 box, I recommend removing it and replacing it with a later-style 800- or 600-series box, which is sometimes called the 670 box. The 605 boxes are worthless for any performance applications, but 1955–1957 Chevy owners often buy them because they are an easy retrofit on those cars.

A stock 800-series power steering box is not be confused with the 605 box that came on some early models. The 800 has a four-bolt cover (shown) while the 605 has a round cover retained by a snap ring.

AGR makes an aftermarket 800 box, and it's an affordable high-performance option. It's far better than having your existing box professionally rebuilt.

All other G-Body models had the 800 box, which has the rectangular four-bolt cover. It was introduced on the very first A-Body models in 1964 and works beautifully. It is very strong, is available in ratios to suit anyone from a road racer to a grandmother, and it is very reliable. Circle track racers use them almost exclusively, and because of this, there is excellent aftermarket support. These boxes were in service in nearly every GM vehicle from the mid 1960s into the 1990s. Early models have the same flare-style fittings as the early 605 boxes, but later ones (1982 or so up) have the more common O-ring style fittings.

Lee, Sweet, and Detroit Speed & Engineering (DSE) all offer adapter fittings to change older boxes to the superior O-ring seal, as well as adapters to go to -6 AN if desired.

Upgrades

For an easy upgrade, you can install the steering boxes found on Buick Grand Nationals, Olds 4-4-2s, and Monte Carlo SSs on other G-Body models. These high-performance steering boxes use a quicker ratio (12.7:1) than most other models. Because they are also 800 boxes, they are a direct bolt-on.

Other G-Body models equipped with the F41 suspension option were fitted with the same steering box. Look for the YA code on the bottom cover of the box. Remanufactured ones from NAPA (PN NSP-88-277101) are available for about $144. Lower performance models with the 605 box used a smaller diameter Pitman arm, so you also need the later-style Pitman arm, which isn't available new at this time.

The S10 ZQ8 boxes are also popular, mainly because they have quicker ratios and are more likely to be found in good condition (used) than the older G-Body boxes. Boxes from Jeep Grand Cherokee models (codes AL, JH, KD, WK) can also be used, but require the use of the original G-Body Pitman arm. The stops increase the turning radius slightly, but could result in tire rub depending on the size of the front tires. Boxes from third-generation F-Bodies, particularly IROC and GTA models, have good (12.7:1) ratios, but the internal stops in these boxes are different and decrease the turning radius substantially. I don't recommend these boxes for this application.

The aftermarket provides plenty of support for the 800 box. If you want to have a box built with a specific ratio, even one that wasn't offered stock, steering experts, such as Lee or Sweet Manufacturing, can blueprint and tailor the steering ratio to your exact requirements. These boxes are still very popular in circle track racing and are a viable alternative to newer 600-series boxes.

The 670 box has a rack-and-pinion feel without the problems of a rack swap. Some call it a 600 series. Several aftermarket sources, including Detroit Speed & Engineering, offer the 670 box. (Photo Courtesy Detroit Speed & Engineering)

The 600-series box (also known as the 670 box) is also a popular choice for high-performance G-Bodies. Similar in appearance to the 800 box, the 600 box has a smaller cover. The 600 outperforms the 800 steering gear because it has a rack-and-pinion spool valve, which gives excellent response and road feel. This box easily replaces the 605 or 800 that was originally installed. Unfortunately, in salvage yards these boxes are only found on trucks, and they have stops and slower ratios that aren't desirable for a swap.

Fortunately, companies, such as DSE, Borgeson, and others, make versions specifically tailored for various applications. While these boxes are more expensive than the typical aftermarket 800 box, they are a beneficial upgrade because they provide precise steering control and excellent feedback. They also provide steering feel that's comparable to a new car.

One thing to keep in mind, though, is that most headers are designed around the 800 box, and the slightly different dimensions of this box may cause fitment issues.

Manual Boxes

If you want to remove the power steering for weight savings or simplicity, a manual steering box is an easy bolt-on modification accomplished with factory parts.

Racers and street rodders alike have used the traditional manual Chevy Vega steering box for decades. The Vega box is very compact and lightweight. It also easily bolts to G-Bodies, most GM vehicles, and many other cars. Originals can be difficult to find but plenty of aftermarket sources are available for this box, so availability isn't its biggest problem. However, it's too lightweight for large, full-framed cars, and durability may be an issue. A compatible stock Pitman arm isn't available because it was never installed in a G-Body from the factory, so a custom Pitman arm from a steering box company must be purchased. To properly fit the Pitman arm to your particular car, it needs to be heated and bent to fit.

As previously mentioned, S10 truck front suspension components largely interchange with G-Body components; steering components interchange as well. Fortunately, many 1982-and-newer S10s with the 2.5-liter, four-cylinder engine were equipped with a manual steering box. They have a good feel and bolt to the G-Body frame with no modifications.

In some cases, you can use an S10 Pitman arm, but if it is different than your G-Body Pitman arm, use your old G-Body arm. The S10 box is a little heavier than a Vega box, but in this case, that's good because it's better suited for high-performance applications. However, don't expect to carve corners with this box because the ratio is a little slow. But for a drag-oriented vehicle with skinny front tires, it's just about perfect.

Rack-and-Pinion Conversion

Rack-and-pinion (often simply called "rack") steering has many advantages over a recirculating-ball steering, including improved feel, ease of packaging, lighter weight, often tighter turning radius, and fewer moving parts. Practically every

Street rodders hold this Vega steering box in high regard because of its small size and ease of adaptation to a 1930s chassis (which is typically very narrow). It's an option for your G-Body, but it isn't the best choice.

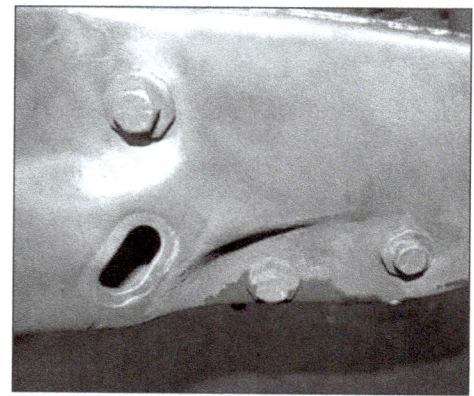

Three bolts hold the steering box to the frame rail. To remove the box, you disconnect the coupler and steering linkage, and of course, the steering hoses.

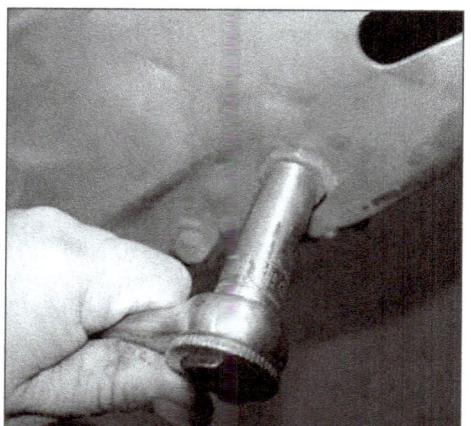

Properly supporting the box before removing the last bolt is very important. Use caution, as the steering box is heavy.

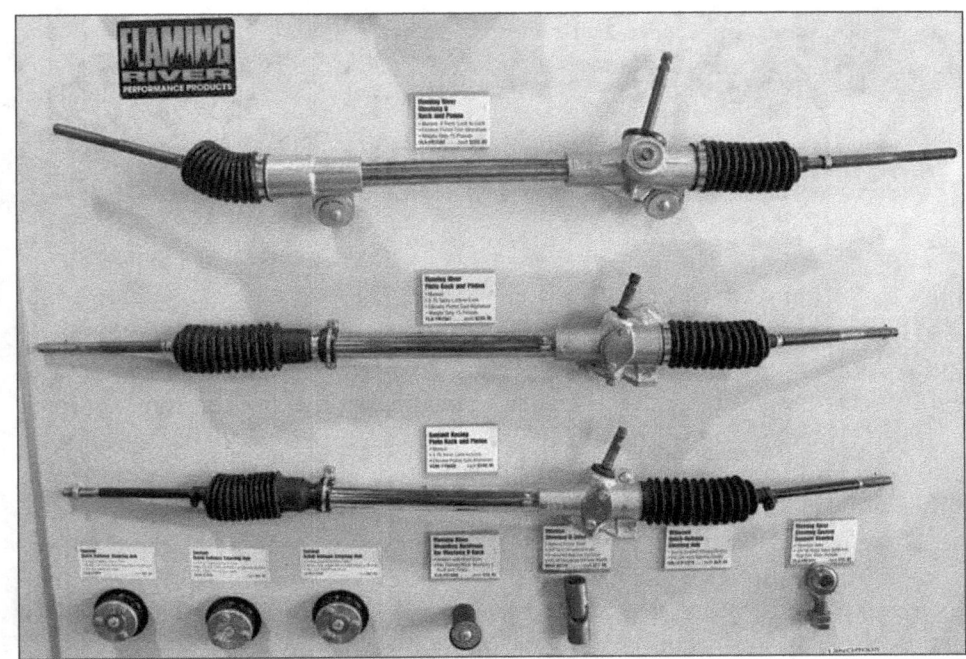

Aftermarket manual rack-and-pinion units are typically based on one of two designs: Pinto or Mustang II. The Pinto is better suited for lighter-weight applications.

automaker has embraced the rack because it's a better steering system than the recirculating ball type.

However, when retrofitting a typical power rack-and-pinion into a car, ease of packaging and tighter turning radius rarely apply. When performing a rack conversion, packaging becomes an issue and often a conversion requires modifying the driver-side header for clearance. In some cases, particularly with an LS-series engine swap, the oil pan must be modified as well.

A rack steering system typically saves weight over a recirculating ball system, but when a stronger, stiffer crossmember must be installed, the weight savings is negated. In addition, the turning radius is severely reduced because a rack requires a much longer steering arm than does a car with a conventional box. Bump steer is often a side effect of a rack conversion, too, which defeats the purpose.

If you are building a chassis from scratch or buying one from a very capable company, such as Schwartz Performance or Art Morrison Enterprises, any of these shortcomings can be overcome because the chassis designer is working with a clean sheet of paper and doesn't have to make any compromises. For the rest of us, however, the old reliable steering box is often the way to go. After all of the following steering upgrades have been installed, you should notice an immediate improvement in steering response.

Manuals

A manual rack-and-pinion is a good option for a drag car that doesn't require a tight turning radius and minimized bump steer. The Ford Pinto manual rack is the most popular unit for this type of conversion. Wild Rides offers a bolt-in conversion kit, but some welding is recommended for the brackets.

Manual racks are as close as the local parts store, or from vendors such as Flaming River, Morrison, or ChassisWorks. Mounting brackets, steering shafts and joints, tie rods, and in some cases support bearings are also needed. Most use a heim-joint end for this type of application because it may be difficult to find an outer tie rod that is compatible with your stock steering arms. The tie rods can often have metric sizing, different taper on the tie rod stud, etc.

Couplers

The stock steering system uses a collapsible shaft and a rubber coupling. Known as a steering coupler (or rag joint), the rubber in the coupling is reinforced. When in new condition, it does a good job of connecting the box to the steering shaft, and it also reduces the vibration that can be transmitted by power steering systems. When play or slop develops in the steering, often the rag joint has worn out and needs to be replaced. Heat, exposure to fluids, and time severely weaken it. When it

Borgeson makes this factory-style steering coupler, also known as a rag joint.

excessively flexes, the steering response is vague and sloppy.

To check the condition of the joint, have a friend get behind the steering wheel and turn the ignition on to unlock the steering column. As your friend turns the steering wheel, watch the joint from under the hood. Determine how much steering wheel effort is applied before the coupler begins to move. If an enormous amount of effort is required, the rubber has broken down and the coupler needs to be replaced. If a moderate amount of steering force is applied and the steering begins to react, the rubber has retained its strength and it does not need to be replaced. Unfortunately, most original couplers fail this test.

Replacements are readily available, but very often, the replacements aren't as well made as the originals. For a restoration, the original coupler must be restored; there is no other way to go. But if you're not concerned with an original appearance, there are other solutions.

Steering Joints

Traditional hot rodders connect the steering box or rack to the column using a high-quality splined (Double-D) shafts with universal joints. This method often resolves difficult routing situations, such as rack-conversion, large-tube headers on a

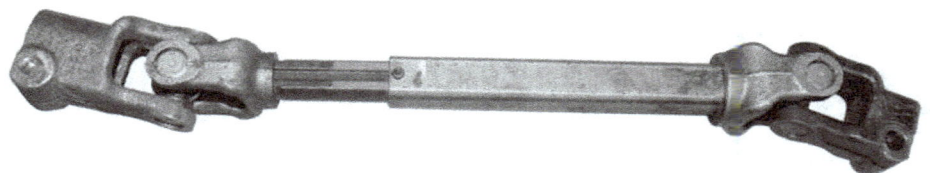

The Jeep steering shaft is found on many Cherokee and Grand Cherokee models, and fits G-Bodies as well.

big-block Chevy swap. In addition, it also leaves more room for error. Much guesswork isn't involved with a typical two-joint system because it either connects the two points and operates without binding or it doesn't. Sometimes you need a third joint to clear an obstruction, but that third joint allows an unsupported steering shaft to travel in an arc, and that is unacceptable. In a case such as this, it is necessary to use a support bearing. The support bearing is basically a heim joint that is solidly mounted to the frame (it's installed in a threaded tube that is welded to the frame rail), and the shaft passes through the opening in the joint (the bearing end). Most aftermarket steering joints have several set screws. They need to be coated with a thread-locking compound and properly torqued down. Some people weld the joints

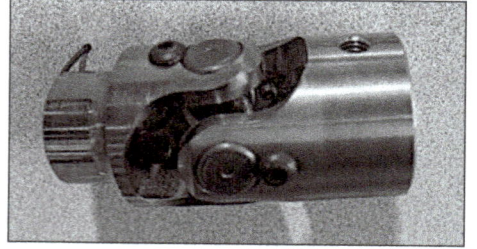

Use an aftermarket universal joint, such as this one from Flaming River, to build your own steering shaft.

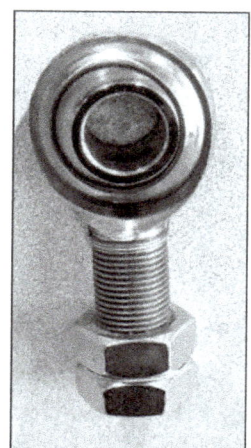

The support joint is needed to support steering shafts when more than two universal joints are used. This isn't typically necessary on a G-Body, but depending on the complexity of your headers (such as on a drag car with a big-block and very large tubes), it can solve some problems.

to the shaft, but this can cause problems. The heat generated from properly welding them together causes the bearings to bind, the grease to melt, and function to be seriously affected. In turn, it causes major reliability and safety issues later. I don't recommend welded shafts for any street-driven vehicle. However, they are solid steel, and won't collapse like factory steering shafts.

Steering Shafts

Late-model GM steering shafts can be retrofitted to G-Body cars as an easier, safer, and less expensive upgrade to the steering shaft. Many late-model vehicles used a factory collapsible Double-D shaft with universal joints, which is very similar to the aftermarket joints mentioned previously. Chevy Astro/GMC Safari vans, Jeep Grand Cherokees, and 1988–1996 Chevy 1500 trucks were fitted with these steering shafts.

For all of the above mentioned shafts, the spline count is compatible with original steering columns and all popular boxes. The shaft length on the above vehicles is close enough that it can be collapsed slightly and installed in a G-Body. Simply press it against the floor, and push until the proper clearance is obtained. This is the system I use in my own cars, and it's readily available in salvage yards at an affordable price.

Here are the general steps to replace an original steering shaft:

HANDLING ON RAILS

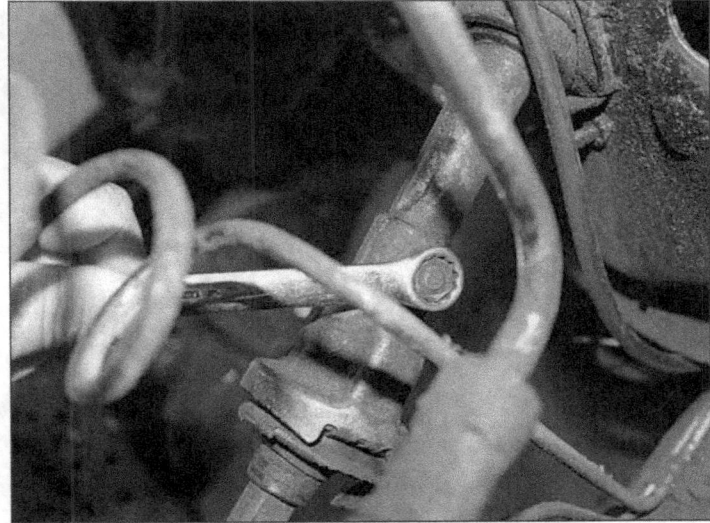

1 To remove the original steering shaft, first loosen the nut securing the top of the shaft to the steering column.

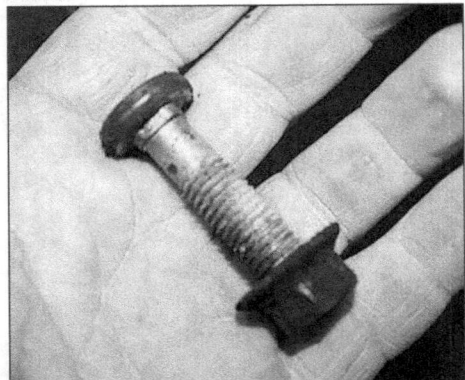

2 Note the factory-applied thread locker on the bolt; this needs to be cleaned off, and fresh thread locker applied before reinstallation.

3 To expose the steering coupler (rag joint), peel back the plastic covering if it still exists. Also remove the bolt at the steering box connection.

4 Even if the rubber discs that make up the coupler appear to be in good shape (not oil-soaked or distorted), a solid steering shaft with U-joints makes a great improvement in steering feel.

5 A Detroit Speed & Engineering 600 series box will replace this 800 during final assembly, but the old box remains in place during mock up.

GM G-BODY PERFORMANCE UPGRADES 1978–1987

6 To remove the original bolt, use a deep-well socket and a long extension because this is difficult to reach in an assembled car.

7 A little persuasion with a pry bar is often needed to get the steering coupler off the splined steering box shaft. Once loose, it's easy to remove the coupler.

8 With the shaft removed, the end of the steering column is visible. Notice this one is a bit nicked up, so we'll de-burr it with a file before reassembly.

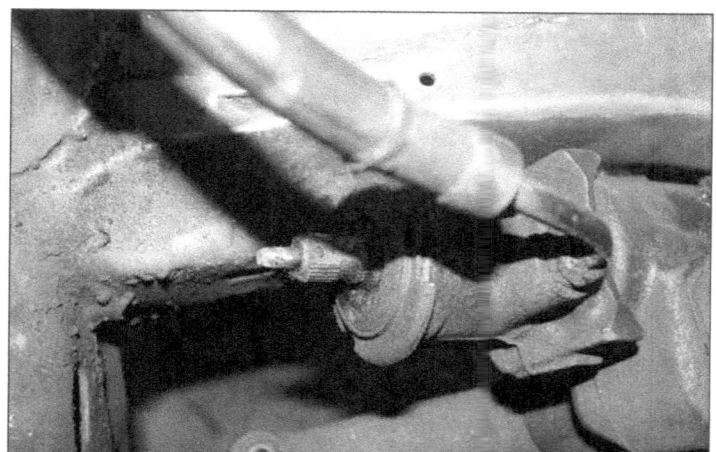

9 The steering box splines are in very good shape, and are the same size as the ones on the Jeep steering shaft.

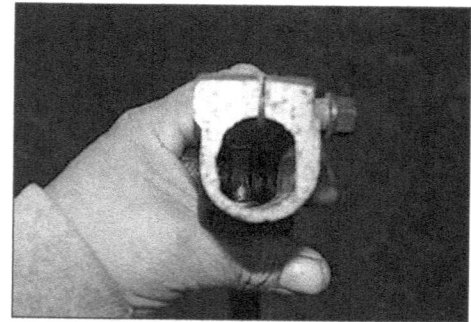

10 Some Jeep shafts use a sheet-metal joint, but this one is cast. You often need a small pry bar to get it over your original splines. Don't use too much force, though, or it will break.

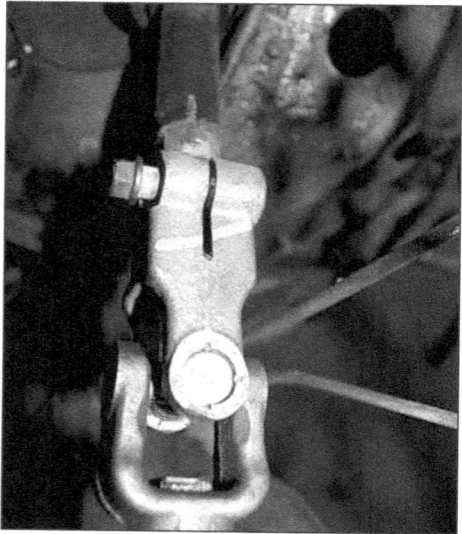

11 Once the splined end of the shaft has been installed, slide the other end over the steering column shaft. You may have to collapse the shaft slightly when doing this.

With that done, insert the bolts (with medium-strength thread locker) and tighten them.

My local supplier built this braided -6 AN power steering hose that's ideal for a variety of high-performance builds; a pair costs $50 to $60.

Power Steering Hose

Power steering hoses must be compatible with the steering box. People usually don't think about the power steering hoses unless they are leaking.

General Motors vehicles into the late 1970s used a flare-style fitting, which includes some of the late A- and early G-Bodies. As long as you are using a stock power steering box and pump, there is no reason to change. Most aftermarket steering box suppliers, as well as power steering pump manufacturers, can provide upgraded parts that work with your existing style of hose.

A steering box with O-ring ports for the power steering hoses is fitted to most of the vehicles in this book. This style is less prone to leak. Also auto parts stores are much more likely to stock replacement parts. When performing an engine swap to a later-style engine, such as an LT1 or any of the LS engines, this is helpful because they already have O-ring hoses. You can use off-the-shelf power steering hoses from another factory application, rather than using custom hoses from a hydraulic hose shop.

A well-intentioned car enthusiast sometimes builds his own braided stainless hoses with aluminum AN fittings for his steering box and pump, which offers a more custom look and perceived quality. Many think this is a step up in quality and performance because this type of hose is used on aircraft and race cars, so it must be better, right? Wrong. The typical braided stainless hoses sold by local speed shops don't have near the burst resistance of a standard power steering hose. The aluminum fittings don't withstand as much pressure and therefore aren't suitable.

Most manufacturers sell their own steering-specific hose, but it's often an ugly color or has some type of woven covering. Steel fittings are a requirement for the pressures a power steering system generates under load.

The solution is as close as a local hydraulic hose shop. Although their primary focus is industrial and heavy equipment, they can make AN-style hoses to your specifications. They have the same braided look that car guys love, but use a Teflon inner hose and steel-plated fittings for increased durability. If you are going this route, be sure to measure the hose carefully and be careful of sharp bends.

The power steering grade of hose is very stiff. Also, make note of the direction of any 45- or 90-degree fittings to be used. Unlike the swivel-type fittings used for plumbing fuel systems, these fittings are swedged on and do not turn. Cost is more than standard, off-the-shelf hoses, but not considerably more. I made my last pair for $75, and I used it to connect a Corvette LS3 Type II power steering pump to an 800 box.

Power Steering Pump

The power steering pump is another component that gets little attention. Most G-Bodies came with a Saginaw-style power steering pump with an integral reservoir, and while not very attractive, they work well for most applications.

Type II

Many owners replace the stock steering pump with a GM Type II steering pump because the stock units are heavy and bulky. The Type II is often used in high-heat applications, such as autocrossing, road racing, and circle track racing because the stock reservoir is soldered; under intense heat, the solder can melt and cause catastrophic failures.

The Type II pump's primary advantage, other than not having a soldered reservoir, is that the

This Type II pump is designed for Corvette LS3 applications, but it's very similar to those used on turbocharged Buicks, newer factory cars, and aftermarket applications. It can be used on LS engine swaps also.

CHAPTER 5

reservoir can be mounted remotely for packaging considerations. The 1986–1987 Buick 3.8 SFI turbo engines are equipped with this pump. Many GM vehicles have used this pump since its introduction with the notable exception of trucks. It works very well, and many OEM and aftermarket reservoir options are available. KRC Power Steering and other manufacturers make heavy-duty versions and complete aftermarket pumps are made of high-grade materials and have improved flow rates.

A stock Saginaw-type power steering pump with an internal reservoir is shown. These work fine for a daily driver, but if you plan on autocrossing or road racing, you may want to upgrade to a remote-reservoir unit. The reservoirs are soldered, and have been known to fail (the solder melts) when overheated in competition.

Salvage yards are a great source of power steering coolers because most newer vehicles have them. In this case, a non-GM cooler was used. This one came from the same Grand Cherokee that donated the steering shaft.

Cooler

Even a Type II pump can benefit from lower fluid temperatures, and a power steering cooler is the most effective way to achieve that. Power steering coolers may be overkill on the average street car, but in a high-performance application, they help maintain power steering pump performance, especially if you plan to autocross, road race, or drift the car. Power steering systems build a lot of heat, enough to melt the soldered joints in an integral reservoir-style pump. Extreme heat cannot only damage components, it also causes the fluid to boil, and results in harder steering. Factory coolers run the gamut in design. Some units are simple, tube-style coolers, which are essentially a loop of hard line where the air moves over the line to cool the fluid. Other coolers have tube-and-fin coolers that mount on the core support or in front of the radiator. Most newer cars have them, and easily adapted units can be found at a local salvage yard or GM parts counter. You can often find a unit from a V-8–powered Cadillac sedan, a truck, or an SUV.

Aftermarket coolers come in three types: tube and fin, stacked plate, or aluminum extrusions. The tube-and-fin style is the least expensive and most commonly used. However, they are the least durable, so extra care in placement of them is needed.

The stacked-plate style is far more durable and less prone to damage. They are the preferred cooler for just about any application.

The extruded-aluminum style that's often advertised as a transmission cooler seems like a good idea on the surface. But they don't cool as well as the other two types, so I avoid them.

Ben Meissner uses this power steering cooler on his *Pumkinator* wagon project. (Photo Courtesy Ben Meissner)

Royal Purple Max EZ High Performance Power Steering Fluid stands up very well to high-performance use, such as autocrossing and road racing.

Power Steering Fluid

Power steering fluid, as brake fluid, is often ignored, but should be flushed and replaced periodically because it degrades over time. I prefer to use synthetic power steering fluid, such as that offered by Royal Purple and others because they have a higher boiling point. This is especially critical for autocross cars because the power steering is put under severe loads.

Derale makes several coolers, including a typical aftermarket tube-and-fin style (top) and a stacked-plate design (bottom). The extruded aluminum type (not shown) is most often seen as a transmission cooler on drag cars.

CHAPTER 6

Hooking Up
HIGH-PERFORMANCE REAR SUSPENSION

As with the front suspension, the rear suspension design and components are a carryover from the earlier A-Body models. While the suspension layout is similar, the geometry is slightly altered so the G-Body control arms aren't interchangeable with the A-Body arms. The suspension is a simple parallel four-link design, with the upper arms being angled to eliminate any need for an axle locating device, such as a Panhard bar or Watts link.

This design provides excellent traction and road holding and provides a huge performance improvement over the leaf spring design that appeared on earlier GM models. As a result, acceleration improved, and it provided far better control of the rear axle in the curves as well. In some models, such as the early Camaro, the leaf spring wraps up or flexes during hard acceleration and exhibits wheel hop and sometimes dramatic loss of control.

Control Arms

Both upper and lower rear control arms are simple U-shaped stampings with large rubber bushings on each end. Inexpensive to manufacture, the factory arms are very weak and the bushings are very compliant, especially once they begin to wear and deteriorate with age. Their main purpose was to locate the axle under the car. In reality they are prone to flexing, easily rust, and a misplaced jack can bend them.

The lower control arms from 1978–1988 G-Bodies are the same,

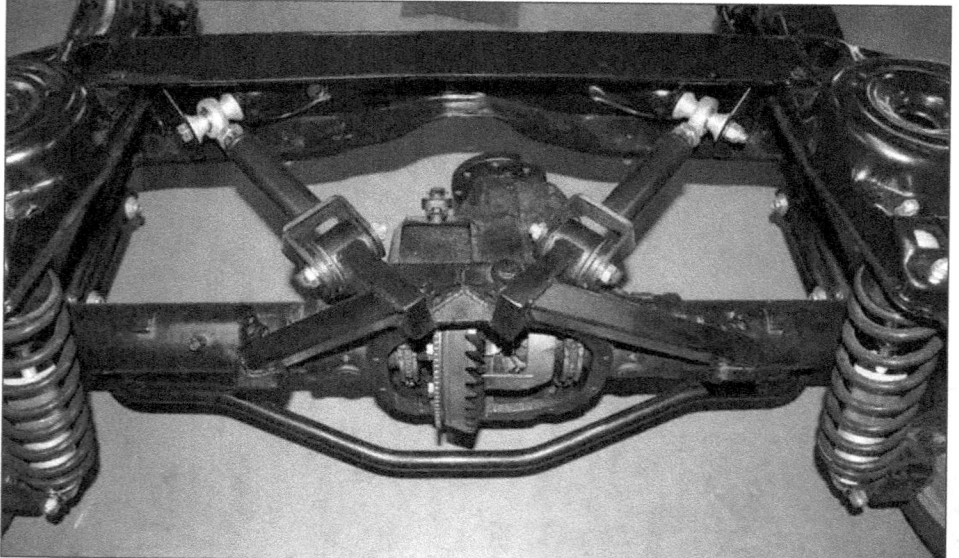

The Sabre, a 1971 Cutlass owned by RnD Fabrication, has a powdercoated stock chassis with a RideTech suspension similar to that of the GNXcess. Note the 8.8-inch rear end conversion. Bulldawg Musclecars may offer this for G-Bodies.

HOOKING UP

and therefore interchange between all models. If you aren't installing a rear sway bar on your G-Body, the 1982–2002 F-Body (Camaro/Firebird) lower control arms are also compatible with the G-Body. They are dimensionally the same, but F-Bodies have the sway bars mounted to the axle, and to the frame, rather than fitting between the lower control arms as on an F-Body.

Bushings

When reconditioning and reinstalling stock control arms, the bushings must be replaced, and it's an easy job in much the same way as the front bushings. A variety of bushings are available, including stock replacement rubber to aftermarket polyurethane bushings. The polyurethane ones are a little easier to install. There is a quick and easy solution if you want to increase stiffness in the original arms, but can't afford or don't want to buy aftermarket pieces.

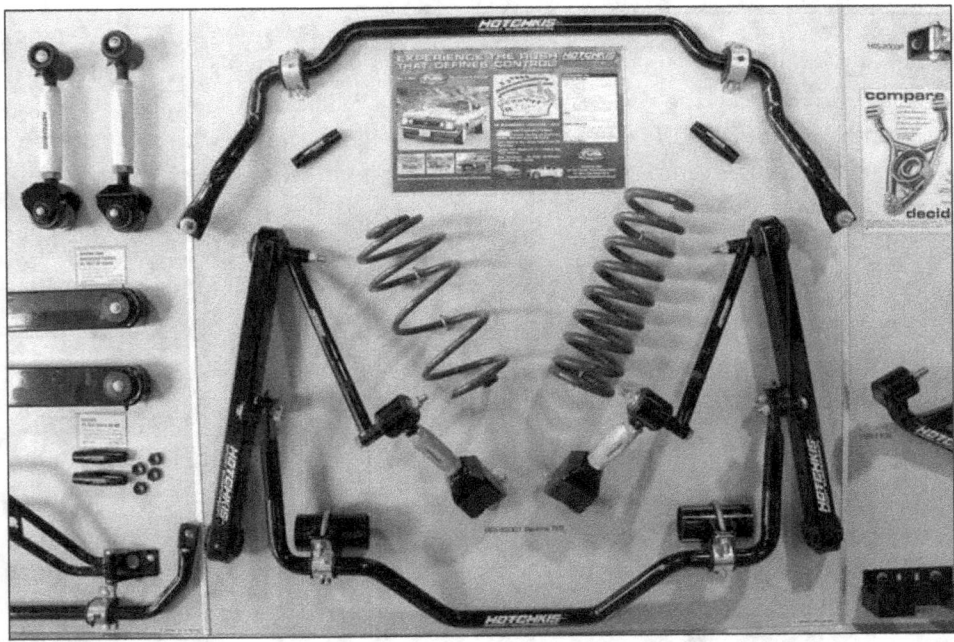

This Hotchkis setup is a good option for those who don't want to replace bushings and box stock arms. The uppers are adjustable for camber to dial in specific handling qualities. Hotchkis and several other vendors offer bracing to tie the lower and upper control arm mounting points together for extra strength. Alternatively, you can build them yourself from steel tubing.

Make a cardboard pattern off of the open channel at the bottom of the arm, and then transfer the pattern to some scrap steel. The 1/8-inch plate works well for this. Cut it out and carefully weld the boxing plate into place. Be sure to alternate the spot welds between different areas of the arm to spread the heat around and avoid warping it. You can box the arms easily and inexpensively. With the right bushings, these arms can work as well as any non-adjustable aftermarket lower control arm. The uppers can be boxed in a similar manner but with less dramatic an improvement.

Most, builders opt to install aftermarket control arms, rather than spend the time to box the OEM arms. Most aftermarket arms carry the OEM suspension geometry of the G-Body, except for the adjustable arms, which are used for fine tuning. Typically constructed of round or rectangular steel tubing instead of stamped steel, these tubular arms are typically stronger than the stamped steel arms. Global West, Hotchkis, DSE,

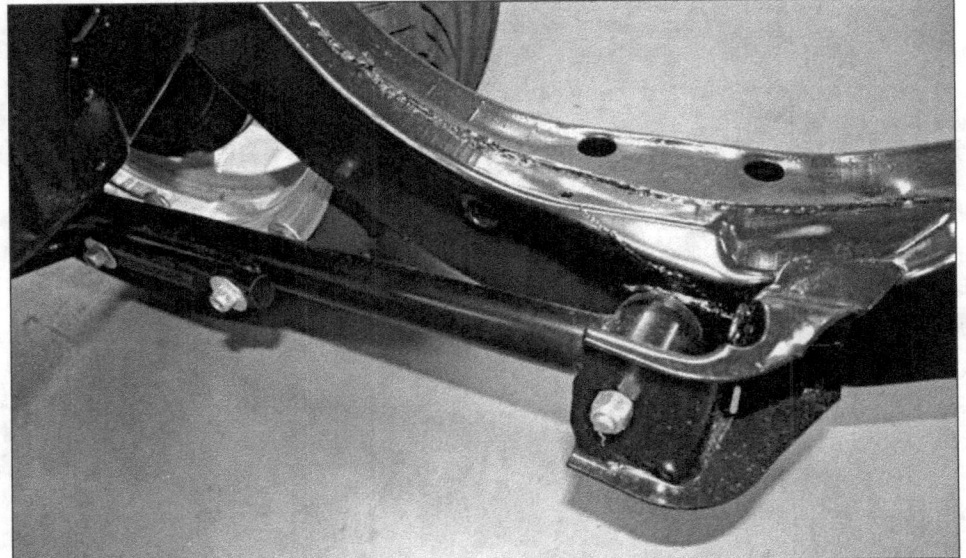

*The RideTech StrongArm lower control arms have been installed on **The Sabre**. They are far stronger than stock and provide a huge leap forward in performance.*

Control Arm Installation

RideTech's StrongArm control arms are made from .219-inch-wall DOM round tubing, and come with a black powdercoat finish. The lower control arms have front and rear urethane bushings; the uppers are equipped with a rod end to allow some adjustment and eliminate binding. The arms bolt up just like the factory ones, using the factory hardware, and have a mounting provision for the sway bar just like the originals. Once both lower arms are installed, the uppers are replaced in the same manner.

Similar to the front, we are installing RideTech parts on the rear suspension of the *GNXcess*. The following series of photos show the rear suspension being fitted to the *GNXcess*. We plan to media blast the entire chassis, add some much needed reinforcement to the rear control arm mounting points, and permanently mate it (and the welded-in 14-point roll cage from Chris Alston Chassis-Works) to the body. Finally, all components will be powdercoated in semi-gloss black.

These are the new RideTech uppers. For now, we have adjusted them to the same length as the stock arms. We'll experiment with other settings later.

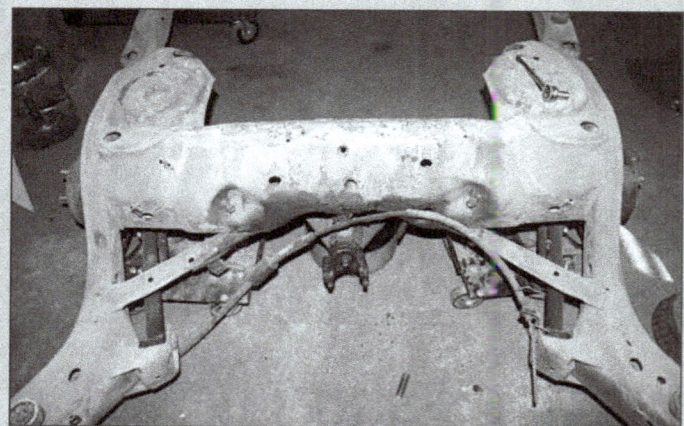

The lower control arms and Musclebar sway bar, also from RideTech, have been installed. The bar installs just like a stock one; it is bolted to the inside edge of the lower control arms, with two bolts per side. There are no end links attaching it to the body, nor any direct attachment to the rear end.

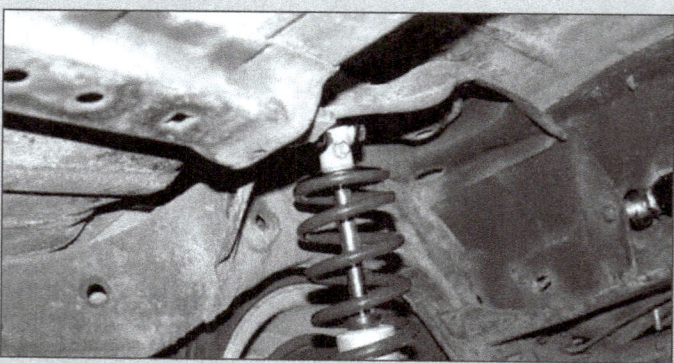

The RideTech coil-overs bolt into the stock shock location, using supplied adapters. The adapters are bolted into the original shock mounting points on the body and on the lower control arm. The coil-over attaches to the adapters. This eliminates any need for fabrication, and makes it a simple bolt-on modification.

The supplied lower brackets bolt up to the rear LCA mounting point. Essentially, the rearmost LCA mounting bolt is inserted through the bracket and the LCA.

ChassisWorks, and at least a dozen others offer tubular arms. Metco Motorsports makes billet aluminum upper (adjustable) and lower control arms. These arms can be ordered with optional instant center modification (ICM) brackets and Delrin bushings (not recommended for street use) in place of the standard Energy Suspension urethane bushings.

Most aftermarket arms come with polyurethane bushings, but some of the higher-end, race-oriented arms have heim joints rather than bushings. These arms with heim joints deflect less than a rubber or polyurethane bushing, but they need to be checked regularly for stretch (take baseline measurements upon installation, and occasionally pull them out and measure; yes, it's a pain on a street car!) and replaced if necessary. They also need to be kept clean, and greased, or they wear out quickly. In addition, the ride is much more harsh than with bushings.

Unless you are removing the rear axle assembly, the easiest way to replace the rear control arms is one at a time. If you are pulling the rear, remove the forward bolts of the control arms only and remove it as a unit.

In my case, I wanted to be able to move the car until the new rear axle assembly had been rebuilt, so I removed the arms one at a time. I sprayed down 25-plus-year-old suspension bolts with good penetrating oil, such as Aero Kroil, and let it soak in for at least a few minutes.

Once that is done, make sure the car is firmly supported, remove the two bolts per side holding the sway bar to the lower control arms, and set them aside. You can use air tools, if you have them, but a ratchet and a box wrench works if you don't (you need both, since the bolts will just spin otherwise). Next, remove the front and rear bolts and the arm.

Sway Bars

The rear sway bar found on G-Body models is a very simple design. It merely connects to the lower control arms using two bolts per side, and therefore it has no end links or other hardware. This inexpensive type of sway bar doesn't control rear suspension forces as well as a more conventional style of rear sway bar with connections directly to the rear end housing and end links connected to the frame rails. As with the front bar, many different sizes are available that interchange among G-Body models, so you can find a very well matched bar using only factory parts.

If you can't find the right sway bar in the salvage yard for your particular setup, the aftermarket offers several versions. Most are designed to work with the same manufacturer's suspension package, so there isn't necessarily "one size fits all." If custom building a suspension, use a proven system from one manufacturer or choose a vendor that carries several sway bars and can make unbiased recommendations.

My RideTech Musclebar is similar to the stock design, and it attaches in the same manner. While many manufacturers offer a similar bar, this one is well matched to the front bar and the spring rates of the RideTech coil-over shocks. As with all RideTech suspension parts, it is finished in a high-quality, semi-gloss black powdercoating.

As for the front suspension, a variety of companies offer spline-type sway bars. Virtually all of them are universal, but they require some degree of fabrication and welding of mounting brackets, so they are somewhat beyond the scope of this book. For most purposes, they are overkill. However, the more drag-oriented bars work very well for their intended purpose and can significantly lower ETs in high-powered cars.

Shocks

As with the front, the rear shocks are often overlooked as a source of improved performance, but a few changes can be dramatic. For general, all-around performance, Koni or Bilstein B6 HD shocks (PN 24-009294) are a large improvement over any of the typical "parts store" shock brands, yet they are still reasonably priced.

Air Bag Suspension

The air lift bag is another popular modification for any vehicle with rear coil springs. Essentially it's an inflatable rubber bladder that is installed inside the spring, which can be inflated to preload the suspension. Originally it was designed to help carry heavier loads or for increased traction in icy weather, but hot rodders soon realized that it was also well suited for drag racing.

While some only run the bag in the right rear spring, some run one in each spring and use an independent air line for each bag, so that it can be fine tuned. My late friend Tom Gerrard ran two on his 1983 Malibu wagon, and said they helped h 60-foot times considerably. He us them on all of his G-Body proje over the years.

Air bag suspensions were developed for tractor trailers.

CHAPTER 6

Doug Lutes' *Sic Monte*

This is a competitive, owner-built G-Body autocross car. It uses the best aftermarket parts, and through trail and error, Doug Lutes built a very quick car that has proven itself in competition. He often competes with the car at American Streetcar series events.

Shown here is the rear suspension of Doug Lutes' Sic Monte. The car has a variety of chassis upgrades that make handling comparable to many other high-performance G-Body cars. (Photo Courtesy Doug Lutes)

Lutes recently replaced his Belltech rear springs and ChassisWorks Vari-Shocks with RideTech coil-overs. While the existing parts were very capable, the new setup allows far more adjustability, which allows him to further dial in his suspension. (Photo Courtesy Doug Lutes)

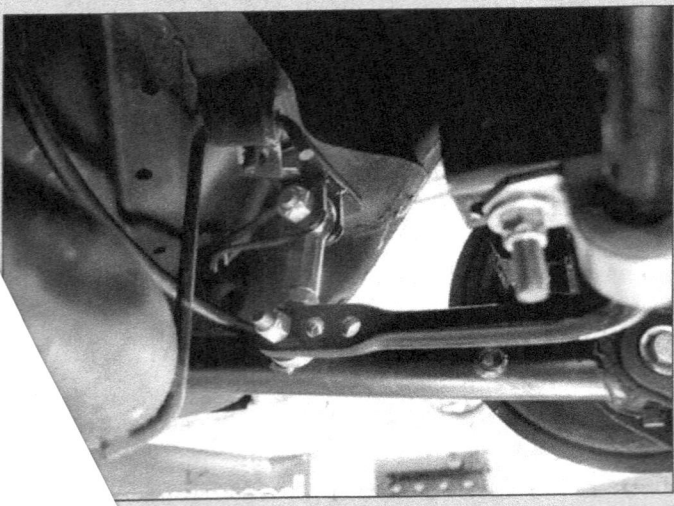

a Spohn adjustable rear sway bar. the end link to the hole closer to the joint ustment is made through another hole. y Doug Lutes)

Currie Enterprises' Curr-Trac upper and lower control arms have also been installed. (Photo Courtesy Doug Lutes)

HOOKING UP

RideTech ShockWaves combine the versatility of a coil-over shock and an air spring. Pictured is one of the new Select Series ShockWaves.

Here is a pair of rear Shockwaves. They bolt on just like a coil-over shock.

The inner workings of a RideTech coil-over shock are depicted here. In practice, a well-matched coil spring and conventional shock performs as well as a coil-over, but the coil-over allows for easier adjustment and spring changes to suit any usage.

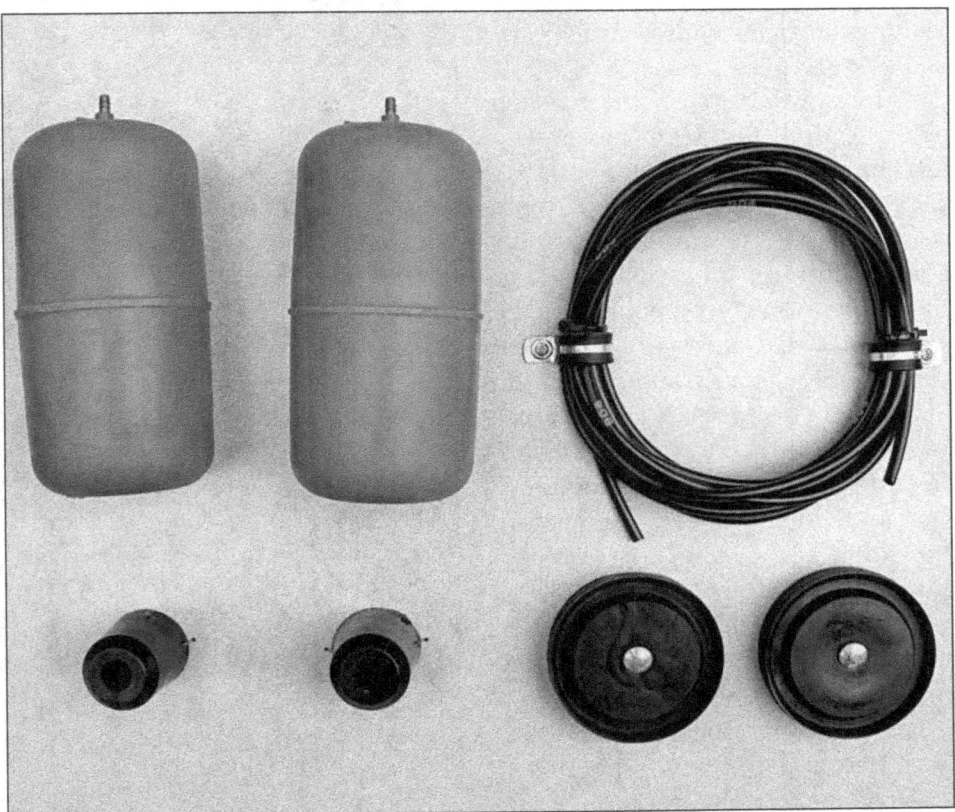

Air lift bags, originally intended for carrying heavy loads in coil-spring vehicles, are also great for preloading the suspension for a better launch. This is a good, low-buck method for increasing traction, resulting in better 60-foot times.

heavy rubber bags became popular with the mini truck crowd as a replacement for springs. This enabled them to get a much lower ride than possible with any other suspension, when used in conjunction with onboard air compressors and special valving. They allowed users to slam the vehicle to the ground at rest, and then raise it to a more reasonable height while driving.

Early systems were often poorly engineered, and garnered a reputation for poor ride and handling. Many were plagued with other problems, such as leaky, substandard valves and poor mounting of the bags or air lines. Although very durable, the air bags can't come into contact with sharp metal (such as inside a front spring pocket on a double A-arm suspension). Also, the air lines must be routed away from sharp edges and exhaust heat.

Modern air bag suspensions, such as those from RideTech, have taken the concept from a car show novelty to a popular and versatile suspension system. Like any conventional spring, nothing inherent in the air bag suspension gives a poor ride. The same rules that apply to coil or leaf springs apply to air bag suspensions.

Spring rate is the key, and the rate is variable, with the addition or subtraction of air pressure, so the setup can be tailored for any application, including autocrossing or road racing. Packaging is one of the problems with air suspension, particularly in the front. You need enough room to install the suspension parts and typical air bag suspensions parts are larger than traditional coil-over shocks.

In some models, space is at a premium and air bags and shock absorbers compete for the same real estate. RideTech's ShockWave has a compact design and solved the problem. It is essentially a billet coil-over shock that has an air bag built around it. A variety of available mounting brackets make it an easy bolt-on modification for most cars, including the G-Body.

In fact, the same RideTech StrongArm suspension components used on the *GNXcess* can also be used with ShockWaves, rather than coil-overs. Electronic controls have also enabled the user to really dial in an air system and use it to its full potential.

The only real drawbacks to air suspension are the weight of the components (compressor, valves, lines, etc.) themselves, and the increased cost over other systems.

The RideTech airpod provides a compact, clean-looking solution to air-spring plumbing.

This is a very nicely detailed A-Body chassis with RideTech StrongArms and ShockWaves.

CHAPTER 7

BUILDING A STRONG FOUNDATION
CHASSIS AND FRAME UPGRADES

Based very closely on the A-Body chassis that was first released in 1964, the G-Body chassis is inherently weak. Built with low cost and light weight in mind, the main rails use thin-gauge steel, and the ladder-style rails have an open channel. Some models came with additional, bolt-on tubular steel bracing that helped, but by themselves aren't enough.

Typically, when the body is mounted to the chassis, the two combined structures of a full-frame car provide considerably greater strength and rigidity. However, in the case of the G-Body, weight-saving measures that were incorporated into its design resulted in a body structure that is also very weak.

The cowl structure, which typically adds considerable strength to the front of a vehicle, is virtually non-existent, particularly on the passenger side of the vehicle where the HVAC system bolts into the firewall. Instead of a solid steel dash structure that is welded in to tie things together, the dash is vinyl and plastic over a thin piece of steel. There is also very little rear seat area bracing and a very weak roof structure.

You have several options for increasing frame strength including boxing the rails, installing a roll bar, and adding crossmembers.

Separate Frame from Body

Boxing the frame rails is one of the most time-tested methods of increasing chassis rigidity, which was done since the earliest days of hot

The entire G-Body stock chassis is shown. When properly boxed and reinforced, it provides excellent performance. The plan is to box and modify the frame of the Pumkinator, *so removing the body was the easiest way to accomplish the work. Bolt-ons and body-mount replacements are easier with the body off the frame. (Photo Courtesy Ben Meissner)*

GM G-BODY PERFORMANCE UPGRADES 1978–1987

CHAPTER 7

Ben Meissner, a fabricator for Bulldawg Musclecars by day and an artist by night at his own company, Street Rod Designs, owns the the Pumkinator, a Cutlass Cruiser wagon. It is under construction, but fully functional. The car houses a Chevy 406; an LS swap is planned for the future. (Photo Courtesy Ben Meissner)

Meissner removed the body using a chain hoist, but if you've stripped it down like this, a few strong friends can get the job done. That is how we did it on the GNXcess. (Photo Courtesy Ben Meissner)

Meissner notched the rear rails for tire clearance, boxed the frame rails in the front and rear, and then painted it. The body cart in the background is very handy during projects like this. You can build one inexpensively and easily with steel, or even wood. (Photo Courtesy Ben Meissner)

rodding. The G-Body benefits greatly from this modification, but you won't commonly find a G-Body with a boxed frame. Due to the construction of the chassis and the way it is tucked up into the rockers, the body must be removed from the frame, so you have enough access to the top of the frame rails to properly weld in the plate metal and box them.

Welding sheet-metal panels onto a frame is intimidating to some, especially those who have never taken a vehicle off its frame before. A full-service restoration shop does this all the time, and it isn't all that difficult, even without a lift.

For example, in rebuilding the 1983 Regal T-Type *GNXcess*, taking the extra step for frame boxing is a no-brainer. The first thing I did was to mock up an LS-series engine and T-56 manual transmission, along with the full RideTech coil-over suspension.

In preparation for chassis work, the entire car was gutted, which means all brake and fuel lines, steering, and suspension components were removed. At this stage, the body is bare and the frame is fully stripped. There's nothing holding the body to the chassis other than the body mounts themselves.

Here are the general steps to follow:

BUILDING A STRONG FOUNDATION

1 If the body mount bolts are stubborn, spray each of them with a good penetrating oil before starting. Once all the bolts are sprayed, carefully remove them with a breaker bar and a socket. Don't use an impact wrench for this, to avoid breakage. The stock body mounts will be replaced with solid steel ones, which completely eliminates deflection and adds stiffness. In addition, the ride should be no more harsh than the typical unit-body car.

2 Once the bolts are removed, the body can be separated from the chassis. This can be done easily with a lift. If you don't have one, the body can be lifted off the chassis by hand (you need four or five moderately strong friends).

3 Slide in 4x4 posts at the front and rear to give you a safe jacking point.

4 Next, jack up the 4x4s from both sides of the car using floor jacks, and then slide in jack stands under the 4x4s to support the body. The frame is then rolled out from under the body.

Once the frame is out, you often find quite a bit of dirt and surface rust on the frame. This can be sandblasted or removed with grinders, but in any case, the metal needs to be clean enough for welding.

Go ahead and install the transmission crossmember. Be sure to leave enough room to remove and install it. You don't want it to interfere with your boxing plates. Some companies may offer pre-made boxing plates, but I recommend making them yourself because it is the simplest way to ensure that they properly fit your car.

Frame Rail Boxing

To make your own boxing plates, follow this procedure:

1 Make a template of the frame rail. Carefully cut some poster board or thin cardboard, and tape it into place on the lower part of the frame rail. At the top, hold the board against the edge of the rail, and lightly tap it with a hammer along the entire length, so you have an accurate template of the needed rail. This shows you where to cut the template. Add more board as needed until you have the entire template for that side. Since the rails aren't identical from side to side, repeat the process on the other side.

2 Once you have made the templates, you need some 1/8- or 3/16-inch steel plate to make the boxing plates. Use a plasma cutter to cut the plates and gussets (an oxyacetylene torch or even a cutting wheel can be used though they take more time). If you don't have that equipment, use an angle grinder with a cutting wheel. Certain metal suppliers, such as Metal Supermarket, can shear and brake metal to your specifications. Because the boxing plate isn't perfectly even, you can just buy a portion of a sheet that is large enough to cut them both out yourself.

3 If you want to add strengthening gussets inside the frame, now is a good time to do it, though it isn't necessary in most cases. Simple rectangular plates, cut from the scrap left over from the boxing plates, can be made and welded vertically into the frame rail. Weld them to the outer, upper, and lower parts of the original rail at regular intervals (every 8 to 10 inches is sufficient).

4 Make the boxing plates slightly oversize to allow for perfect fitment to the rail. Grind the edges of the boxing plate after it is cut. Then mock it in place and tack weld it in a few locations. This can be done with either a MIG or TIG welder.

5 Once the plate is tacked in place, alternate between grinding the box plates flush with the rails and stitch welding between the spot welds until the panel is perfectly fitted to the frame rail. Be sure not to apply too much heat and warp any area of the frame. Quench the welding area with water or air if necessary.

6 The rail can then be completely welded, and then ground with a flap wheel to hide any evidence that the rail was ever anything but fully boxed.

Roll Bar and Roll Cage Installation

Roll bars and roll cages serve several purposes. The obvious one is that they protect the driver and occupants in the event of a crash, but they also tie the chassis together for much greater torsional rigidity. That means the chassis and the suspension handle better under higher speeds and greater loads.

Installing a roll bar or roll cage is somewhat similar to adding subframe connectors to a car. With a roll bar, you're tying together the frame and body, and in the process, you're adding rigidity to the chassis. A roll bar typically has between four and eight points of contact with the chassis: two for the main hoop, (sometimes seen with two very short bars that triangulate off the main hoop for extra strength), two rear

CHAPTER 7

The chassis and body work are underway. The roll cage will eventually be installed. One of the most challenging and time-consuming body modifications is installing wheel tubs. You can cut out the stock wheel tubs and fabricate your own inner wheel tubs for extra tire clearance. It takes a lot of skill to cut, shape, and weld inner wheel tubs. (Photo Courtesy Ben Meissner)

Here is Meissner's wagon, with the cage installed. Once again, he carefully planned, measured, cut, and bent the tubes for this roll bar. (Photo Courtesy Ben Meissner)

Meissner's roll bar was installed more for chassis rigidity and autocrossing, rather than any specific racing class, and it's technically not a cage, at least not in the front. He wasn't as concerned about the front downbar positioning. This car is a daily driver so entry and exit were the biggest considerations. (Photo Courtesy Ben Meissner)

downbars to the rear frame rails, and two forward bars that connect the main hoop to the front portion of the frame rails. A bar that has all of these points is considered an eight-point roll bar.

A roll cage has an additional "halo" bar, essentially another hoop attached horizontally to the front of the main hoop. This serves as rollover protection, with two or more bars supporting it at the front and tying it into the frame and forward bars. Roll cages typically have at least eight points, and some have fourteen or more. The cage I am using for the *GNXcess* is a fourteen-point cage kit from Chris Alston ChassisWorks. Before you decide whether to install a bar or cage, determine the application of your car, such as pro-touring, modified street car, road racer, drag racer, or autocrosser. Most forms of racing have very specific requirements that need to be followed to the letter, and these are spelled out in the sanctioning body's rulebook. Get a copy and incorporate it into your plans from the start to avoid having to remove and install the bar or cage a second time. Remember, these rules are for your safety and the safety of other competitors, so don't try to get around them.

I know that most enthusiasts like to do everything themselves. However, if your welding skills aren't very good, seek the services of a professional for all structural welding. A poor welding job isn't safe and won't pass tech inspection.

Our project car is going to be used for all kinds of competition from drag racing to autocrossing and road racing. Ultimately, this car will do some land speed runs as well, so lots of homework must be done to ensure that it can fit the rules of several sanctioning bodies.

BUILDING A STRONG FOUNDATION

The roll bar in Doug Lutes' Monte Carlo SS allows easy access to the driver and passenger seats while still significantly stiffening the chassis. The car is used mostly for autocrossing, so a full cage isn't required.

The bar doesn't attach directly to the floor. Instead, a steel reinforcement plate is installed between the thin sheet metal and the tube to minimize chances of tearing during a rollover. A G-Body is a full-framed car so pass the tube all the way through to the frame rail for maximum protection and chassis rigidity. NHRA and other race sanctioning bodies require this. It's a good idea to plate around the tube where it passes through the floor. (Photo Courtesy Doug Lutes)

Note the pin that retains the harness bar. It can be easily removed to take the rear bar out if desired. Since the car has no backseat and doesn't retain the factory belts, it isn't really necessary, but it can be a nice touch on a less track-oriented car than this one. A swing-out arrangement also works well on the front downbars (if equipped), so that they can be removed, or at least disconnected, for easier driver and passenger entry and exit. (Photo Courtesy Doug Lutes)

Once you have picked the bar or cage configuration needed, the next critical step is to prepare the car for proper installation. Follow these steps:

1 Completely strip out anything that could possibly catch fire during the cage welding process, including carpet, underlayment, sound deadeners, seam sealer, etc.

2 If the glass wasn't removed, protect it from any stray welding or grinding sparks with heavy cardboard or welding blankets.

3 Mount the front seat to the correct location for the primary driver or drivers of the vehicle, so you can determine the location of the main hoop, steering column, and steering wheel. All too often, a builder installs the main hoop and starts welding, with no consideration for the seating position and location of the controls. The result is a car that is uncomfortable or unsafe to drive, if not both.

4 Once you determine the position of the main hoop, mark the floorpan around it and cut out enough metal to give clear, unobstructed access to the top of the frame rail. Unlike with a unit-body chassis, you weld the main hoop directly to the frame rail, not to the floorpan.

5 A plate is provided that is welded in for extra strength, before actually welding in the main hoop. Measure the thickness of the provided plates, as most are going to be spec'd for drag racing. Some other types of racing, land speed in particular, require thicker plates than drag racing. The plates are provided completely flat and may not fit tightly against the frame, so it's best to heat them with a torch and hammer form them to fit as closely as possible.

6 Once you are satisfied with the fit, weld the plates in place. Please note that either MIG or TIG welding suffices for mild steel, but you should use TIG if the tubing is chrome-moly. If you have the option, TIG always provides a nicer looking weld, but MIG can be just as strong. Keep in mind that no grinding or smoothing of welds is permitted if you're going racing.

CHAPTER 7

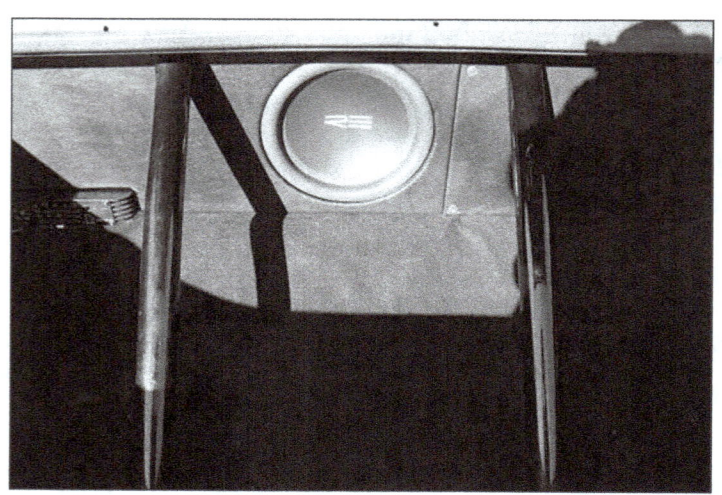

The rear downbars on this Malibu are a little far inward to be attached directly to the frame rails, but may be tied in to an added-on crossmember under the floor.

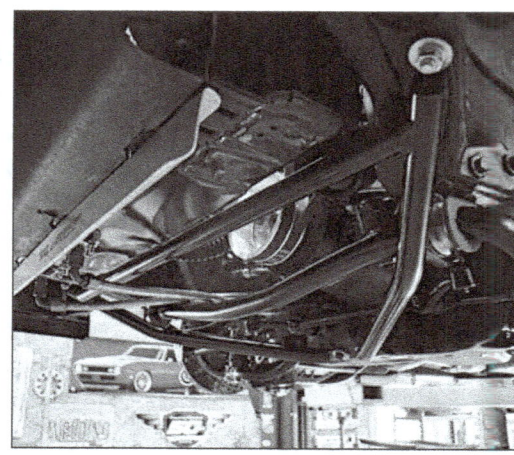

Spohn made this front chassis brace. It is similar in appearance to the factory braces found on many G-Bodies, but offers much greater strength due to the welded construction and larger-OD tubing. Even a factory-style brace is a great improvement, so don't hesitate to grab one from the salvage yard if you don't want to buy or build a new one. (Photo Courtesy Doug Lutes)

7 With the plates in place, ensure that the main hoop fits snugly against them, with no gaps, and that the main hoop is level. In most cases, just carefully and gradually grind or notch the tubing for a tight fit. And "tight fit" is the key. You don't want to fill a sloppy joint with weld because the strength of the joint is compromised.

8 Tack the joint in a few places, and then move to the rear downbars.

9 In some cases, the rear bars pass through the backseat area, but most need to go through the package tray area. This can be easily opened up with a cutoff wheel. I usually just cut out a section from the front of the package tray, run the bar through, then trim the part that I removed for clearance, and weld it back in place. That way, you still have a clean look.

10 Weld the end that attaches to the frame rail in the same way as you welded the main hoop. You often need a tube notcher for the end that welds to the main hoop unless your roll cage supplier notched the tube for you. Use an angle grinder to notch the tube. Specialized tube-notching tools are available for less than $150, and they notch tubes much faster than you can do it manually (with a die grinder). Do the downbars one at a time, tacking them as before, and be sure they are at the same angle and position relative to the frame rails.

11 The front downbars or door bars are more complex to install. For example, if the car uses an armrest, be sure it is bolted in place, and have the window crank or window switches in place so the bar does not obstruct them. If needed, some builders put a slight bend in the bar for clearance while others just notch the armrest.

Some install a swing-out door bar kit for perceived easier access, but it can be difficult to install as well. Its primary advantage is for a street/occasional competition car because it is completely removable for daily driving. If you do this, don't remove the crash bars built into the doors because the extra weight savings from removing them isn't worth risking your life. Most of the remaining bars in a roll cage system install in a similar manner. However, you need to allow enough space to carefully mount any horizontal bars in front of or behind the dash and the front downbars from the halo bar. Very little room is available under the dash of a G-Body, so be sure that the dash goes back into the car with the bars in place and these bars do not interfere with any of the car's other hardware. Stock HVAC systems are most often removed to save space, and even aftermarket ones should be carefully mocked up and fitted.

Similar consideration needs to be given to any bars in the engine compartment because adequate clearance must be allowed for header installation, brake master cylinder/booster clearance, and the stock inner fenders (unless you plan to fabricate new ones).

12 Finish weld all joints and replace any sheet metal that was cut away for frame access once every bar is mocked up into its proper position, and all possible obstructions are accounted for.

Crossmembers

A number of options are available if you don't want a cage or bar, but your chassis must be stiffened for high-performance and/or competition applications. Simple bolt-ons from other G-Bodies are often available. As previously mentioned, some models came with tubular braces in various locations, such as fender-to-radiator support, from the center of the front crossmember to each frame rail, and even X-braces behind the rear seats. The X-braces are found on some Grand Prixes and the ultra-rare GNX. You can find these braces in salvage yards or on eBay. If you have even moderate fabrication skills, you can bend the metal and weld them in.

Ben Meissner, the owner of the *Pumkinator* Cutlass Cruiser wagon, built many of the braces on his car, including a pair that link the rear upper and lower control arm mounting points. An additional rear crossmember, located behind the fuel tank, can increase protection from a rear-end collision, as well as stiffen the chassis considerably. In Ben's case, he used one that was already installed on his project car, and it served as the mount for a trailer hitch.

Similarly, a simple round-tube crossmember in the engine compartment, just ahead of the

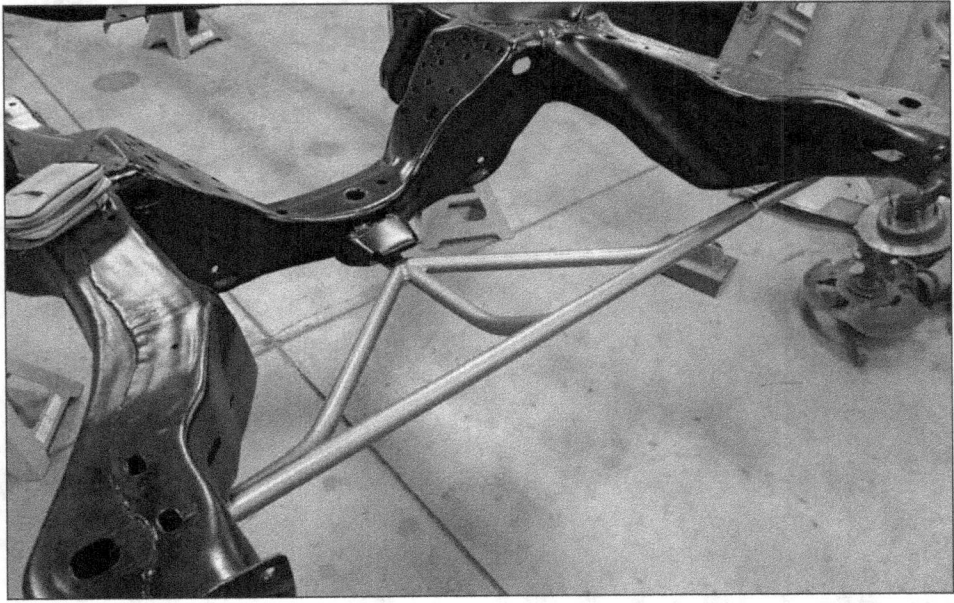
The owner fabricated this orange brace on the Pumkinator. *The G-Body chassis is extremely weak, so additional gusseting and bracing makes a huge difference in torsional rigidity and, ultimately, performance, whether the car is to be caged or not. If you have access to a tubing bender and a welder, you can fabricate many of these types of braces, saving money over commercial offerings. (Photo Courtesy Ben Meissner)*

Some models are equipped from the factory with a simple steel fender brace (which works fine), but this one is made of carbon fiber, and has heim joints for easy adjustability. This is more for the look than any real performance gain over steel parts. They are available from GNS Performance and others. (Photo Courtesy Doug Lutes)

Meissner used a tubing bender and MIG welder to fabricate this upper shock brace. Since this is a very weak area of the chassis, it adds considerable strength. For making templates, a piece of 1/8-inch steel round bar bent to suit works well. The round bar is inexpensive and easier to work with than the tubing used here. Most bending mistakes can be eliminated when using the more expensive seamless 3/4 DOM tubing. (Photo Courtesy Ben Meissner)

CHAPTER 7

Meissner built very simple braces that attach the lower control arm mounting points to the uppers and the crossmember. The crossmember contains the driveshaft and strengthens the lower control arm mounts. All of these bolt-in parts can be easily removed for service. (Photo Courtesy Ben Meissiner)

This underside view is of Lutes' Monte SS, showing a well-thought-out chassis. (Photo Courtesy Doug Lutes)

steering box, can also add considerable strength and prevent movement of the rail under hard steering. IROC Camaros came with a similar brace, called the "wonder bar," for the same reason. It bolted to the front sway bar mounts.

Replacing worn body-mount bushings also helps increase stiffness and improve the ride. Made of rubber, they deteriorate from age and the elements, and eventually begin to fall apart. The bushings themselves are fairly easy to replace, even while the body is on the frame. Many companies offer these bushings in a variety of materials. Stock replacements are rubber, but they are also available in urethane, steel, and aluminum. When the bushings are in good shape, the material doesn't affect ride or stiffness, but urethane and metal last longer. The frame isn't designed to move, so using harder materials, such as urethane and metal, is not a detriment.

However, bushings with a stiffer material composition magnify any suspension problems, and if your G-Body has a stiff suspension, they often produce a harsh ride. Our project car will be fitted with a full cage and the chassis and body will permanently be joined, so I will use more durable steel bushings to maintain integrity.

Here are the steps to follow for replacing body mounts:

1 Separate the frame from the body. This is a good time to clean the frame and the underside.

2 Choose new hardware and the best mounts for your application. You can get suitable hardware through an industrial fastener dealer or buy them in kit form from most restoration and performance suppliers. You need new hardware because often there is rust and corrosion on the threads, and it isn't unusual to break a few during removal.

3 Spray penetrating oil on all threads ahead of time.

4 You can replace the bushings without taking the body off the frame. Just use jack stands or a four-post lift to support the body.

5 Spray penetrating oil on all the body bolts and let it soak in. Loosen all the body mount bolts by a few turns. Don't remove all of them yet because you don't want to throw off the alignment of the body to the chassis. Removing all the bushings at once can do that.

6 Once they are all loose, remove and replace each worn bushing with the new bushings and hardware, alternating from one side to the other.

7 It's a good idea to treat the threads with anti-seize compound to prevent galling.

8 Keep the bolts loose until the last one is in, and then go back and tighten them all.

Wheel Clearance

G-Bodies have decent size wheel wells, but the frame rail intrudes into them and interferes with oversize aftermarket wheels. This used to mostly be an issue with drag cars, but now it's an issue with pro-touring builds and street cars that have larger wheel and tire sizes. In most cases, "tubbing" or "mini-tubbing" isn't needed because the wheel wells are often large enough for the wheel/tire combination. Frame rail interference is the real issue. To remedy this, the frame has to be notched.

Frame Notching

The idea of cutting the frame sounds a little scary, but it really isn't difficult. The frame is fairly thin, and can be easily cut with a 4½-inch grinder or a reciprocating saw. A plasma cutter makes quicker cuts, but a blade or wheel is cleaner. Here are the steps:

1 Determine the clearance needed for your wheels and tires to be sure this operation is sufficient. A clearance of 1/4 inch is the minimum with a typical street tire. A car that is going to have "sticky" race rubber needs more clearance, at least 3/4 inch. This gives room for the tire to "grow" as it heats up.

2 Cut only into the seam, where the frame rails come together. The inner part of the frame is thicker, so you aren't actually cutting away half of the frame.

3 Write down how far forward and how far to the rear the

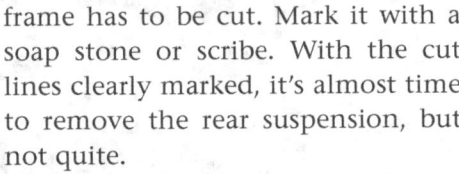

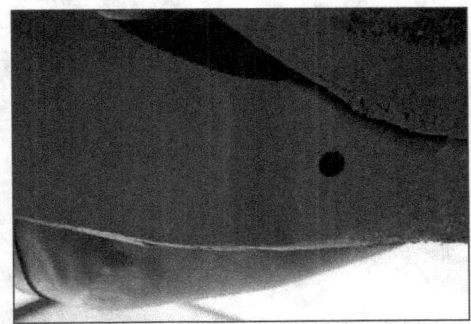

Doug Lutes' Monte SS needed more room for larger tires, so frame notching was in order. This can be done with the body on, but if you're doing a full restoration, having the body off makes it easier. You can see the 1/8-inch plate steel that bridged the gap when the frame was notched. This is the rear section, which gives an idea how far back the notch needs to go. (Photo Courtesy Doug Lutes)

frame has to be cut. Mark it with a soap stone or scribe. With the cut lines clearly marked, it's almost time to remove the rear suspension, but not quite.

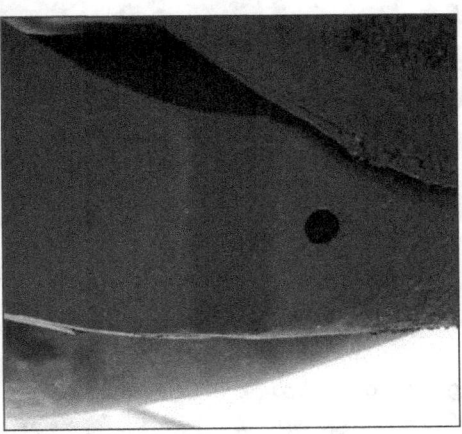

The front section of the notch is visible here. You can get an idea where the frame needs to be notched, but you should still mock up your own wheels and tires to ensure that you have sufficient clearance. (Photo Courtesy Doug Lutes)

This shot of the finished frame notching shows just how clean the modification is. If you didn't know what the stock frame rail looked like, you wouldn't know anything had been done. Fresh black paint and retaining the factory plastic ID tag in the rail add to the stealth effect.

4 Support the rear of the frame rail behind the arch with a floor jack or adjustable jack stand to prevent any movement when you start cutting.

5 Cut a piece of mild-steel tubing of about 1½-inch diameter. Place it between the rails in the area above the rear axle. When determining placement, make sure that it doesn't contact any part of the suspension when it is under full compression.

6 Once you have determined the best location, use a wire brush to clean off the frame rails.

7 Tack weld the crossmember in place and use a square to make sure it is straight.

8 Double-check the clearance and then stitch weld between the tack welds.

The initial cuts have been made on the frame rail. The gauge of the frame is fairly thin so a reciprocating saw with fresh blades and die grinders are easily able to cut the frame rail. A plasma cutter works well also, though a blade or disc gives a finer cut. (Photo Courtesy Ben Meissner)

The frame rail section has been cut out. Some people fabricate a new section from steel plate, or buy a pre-made frame section made for the purpose. Here the original frame section was used. Add some gusseting inside the frame while you have it apart, for added strength. (Photo Courtesy Ben Meissner)

Test fit the frame section before welding. All weld areas must be clean in order to stitch weld the frame together in small sections. The frame rail seems very strong, but still warps if you put too much heat into it. (Photo Courtesy Ben Meissner)

The rail has been welded in, except the end pieces. These were made from steel plate. Good weld penetration is key because this is a major structural area of the car. If your welding skills aren't up to snuff, take the car to a professional. (Photo Courtesy Ben Meissner)

BUILDING A STRONG FOUNDATION

There's more than enough room between the existing Nitto P275/40ZR17 rear tire and the frame rail. The notched frame delivered more clearance than necessary for this combination, but Meissner plans to install larger wheels and tires in the future, so the extra room is needed. (Photo Courtesy Ben Meissner)

Now it's time to remove the rear suspension. Here are the steps:

1 Begin making cuts into the frame rail. I use a thin, 5-inch cutting wheel on a power grinder, because it is easy for me to control and get nice, straight cuts. Take your time, and be sure to wear safety glasses and a face shield. It's no fun pulling metal out of your face with tweezers, or worse yet, having an ophthalmologist cut it out of your eye.

2 Carefully cut along the lines that were previously marked.

3 Make another cut to the inside of the frame rail's weld seam, on the bottom of the rail.

4 Finally, cut along the top seam. Depending on your cutting tool, it may be necessary to loosen the body mounts and raise the body slightly off the frame in the rear to gain clearance.

5 With the rail section cut out, use a grinder to clean up the edges in preparation for welding, using a 40-grit disk on the same power grinder.

6 While not required, it's a good idea to add some gusseting inside the frame, for extra strength (use the same method described in step 3 on page 79 for frame rail boxing).

Next, you need to decide what to use to plate the frame rail back in. I use cold-rolled steel plate and the same pattern-making method described in steps 1 and 2 on page 79 for frame rail boxing. Follow these steps:

1 Use locking sheet-metal clamps (basically Vise-Grips with a C-clamp end) to hold the steel plate in place.

2 Tack weld it into position well, following up with a healthy weld bead. A good MIG welder handles this with ease, or a TIG can be used if you have one available. Either way, alternate when welding from one side to another to evenly distribute the heat and prevent warping the metal. Use air or water to quench the area being welded if necessary.

3 Once the first side is done, repeat on the other side.

Installing Mini Tubs

In some cases, notching the frame doesn't provide enough clearance for the wheel, so installing mini tubs is required. Since this is a complex and challenging project that requires extensive fabrication, many enthusiast builders do not attempt it. Instead, a professional shop completes the fabrication.

To ensure that the wheel well is no longer an obstruction to the wheel/tire, it must be moved inward, so that it is even with the frame rail. Here are the basic steps.

The Pumkinator's mini-tubs can be used as a model if you have a wagon and plan to use it as a family car. It retains the rear seat, so careful measuring was done to be sure the seat would still fit after the modifications.

CHAPTER 7

1 Carefully cut out the existing inner wheel well using an angle grinder with an abrasive cutting wheel.

2 Cut away the support for the trunk hinges.

3 Tack weld the wheel well back in place and add a filler strip of sheet metal (16- to 18-gauge) to fill the gap.

Meissner fabricated most of the tub because it was easier than modifying the stock wheel wells. (Photo Courtesy Ben Meissner)

4 Weld the wheel well in place and use a metal working (body) hammer to fit the contour of the wheel well.

5 Weld the trunk hinge support to the new, wider tub.

6 With all the welding done, grind the wheel well smooth, and paint or coat it to match the rest of the trunk. When properly done, there are no signs of the surgery either inside or outside the car.

Aftermarket Chassis

Buying a full custom chassis provides enhanced performance and it is much more convenient than modifying an OEM chassis. The aftermarket chassis is much stronger than stock, and it has excellent front and rear geometry plus plenty of wheel and tire clearance.

Over the years, several companies made GM F-Body and A-Body bolt-in chassis but not for the G-Body because it wasn't nearly as popular as the F-Body and original A-Body GM cars. A chassis company can custom build a G-Body chassis for you, but typically the chassis doesn't include body mounts, radiator supports, or other needed parts. Most are designed to have the body channeled over them, so a full floor has to be fabricated. This turns into a major fabrication project very quickly, putting it out of reach of all but the wealthiest builders.

Well-known racer Jeff Schwartz of Schwartz Performance offers a complete custom G-Body chassis.

Schwartz Performance chassis are very well known in pro-touring circles and have been built for most muscle car applications. Its latest addition to the line is one for the G-Body. Amazingly, the chassis has 200 percent more torsional stiffness than the original, and is 119 pounds lighter.

Schwartz uses RideTech single- or double-adjustable coil-overs, with several brake options, including those from Wilwood and Baer. This chassis is a direct bolt-in on a G-Body and requires no modifications other than to the gas filler neck. The tank is centered in the new chassis to allow for more tire clearance and mini-tubs, if desired.

Full-floater 9-inch rear axles with a triangulated four-link are standard. A 9-inch-based independent rear suspension is optional. With mini-tubbing, either can accept up to a 345-mm rear tire. Schwartz chassis are available powdercoated, and come pre-assembled, so installation is a snap.

Dan Howe's Monte Carlo received the first Schwartz G-Body chassis. The chassis was designed for a lower stance and delivers a lower center of gravity, which results in improved stability and performance. It has full suspension travel and still has a comfortable ride.

Howe's Monte Carlo SS looks like a professionally built car, but he and his son Josh did nearly all of it themselves. Josh did the body and paint work. They have competed in several events including, the 2012 Optima Streetcar Challenge in Las Vegas.

The six-piston Wilwood brakes stop the car in a hurry. Howe runs XXR wheels and BFG G-Force T/A P275/35ZR18 tires.

Dan and Josh Howe mini-tubbed the car without using a kit. They cut out the stock tubs, moved them inboard, and welded them in place. A filler piece made of 16-gauge steel fills the gap.

His 1982 Cadillac and Ultima GTR have appeared in the *Car Craft* magazine Real Street Eliminator competition. Schwartz Performance is the leading supplier of bolt-in chassis for a number of vehicles. The first car to receive a Schwartz G-Body chassis, a Monte Carlo SS owned by Dan Howe, recently competed in the Optima Street Car Challenge. The 1.3-g performance, increased rigidity, and incredibly lighter weight are all features of the Schwartz chassis. Bulldawg Musclecars also carries these chassis.

Howe runs an LS1 with Lingenfelter internals, and a 4L60E. He's planning to add a paddle shifter.

CHAPTER 8

PUTTING THE POWER DOWN
HIGH-PERFORMANCE DIFFERENTIALS AND AXLES

G-Bodies came with a Salisbury-type gearset and differential loaded from the rear, which was distinctly different from other GM cars, such as the A-Body. Since 1964, the A-Body was equipped with the Hotchkis-style differential that loads through the front. Unlike the earlier axle assemblies, these Salisbury-type gearsets were built with overall weight in mind, as well as the reduced power levels expected from the cars that received them. The C-clip arrangement retains these axles, as on most other post–1964 GM rears.

The G-Body's most common rear end had a 7.5-inch ring gear with a 10-bolt differential. This differential was originally designed for the Chevy Vega, which should tell you that strength wasn't a primary concern. The weak 7.5-inch differential came with 26-spline axles, and the case tended to flex in high-torque applications, particularly with a small 7.5-inch ring gear. Some units came with limited slip, and gearsets as low as 3.73:1 were offered in high-performance models. Later GM models (such as the F-Body and S-series trucks) used a slightly stronger 7⅝-inch ring gear and 28-spline axles.

G-Body rears can be upgraded to use the later 28-spline differentials and larger gears, but aftermarket 28-spline axles are required. C-clip eliminator kits are available but tend to leak.

Bulldawg 8.8 rears install in G-Bodies similar to a stock axle assembly, utilizing any G-Body control arm, even the stock ones. This is done by using a truss that has the correct upper control arm mounts attached, and the correct LCA mounting brackets. The Bulldawg 8.8 is based on the cheap and plentiful Ford Explorer rear-axle assembly. These rears come with an 8.8-inch ring gear, 31-spline axles, and factory disc brakes. While I have shown the A-Body unit, a G-Body version may be available and it may look similar to this.

The 7.5-inch is the most common axle assembly found in G-Bodies. Two vertical bolts on each side of the cover identify it; the more rounded bolt pattern identifies the 8.5-inch. This 7.5-inch carries a TA Performance aluminum cover, Precision 3.73:1 gears, and Auburn limited-slip. (Photo Courtesy Doug Lutes)

This G-Body 8.5-inch axle assembly is strong enough for most street-driven vehicles because of its bigger, stronger pinion and ring gears. This one is installed in a 1987 Grand National. Note that the bolt pattern is round, rather than squared off on each side as is the weaker 7.5-inch. There are many cars making 500 hp or more still running the 8.5 with no issues.

GM 10- or 12-Bolt

General Motors differentials must be properly identified. It's not uncommon to mistake a 10-bolt as a 12-bolt differential. The "10-bolt" designation refers to the number of bolts holding the ring gear to the differential case and not the number of bolts on the removable, stamped-steel rear cover. Coincidentally, both have 10 bolts, but these numbers don't coincide with or represent every GM axle assembly. The Olds "12 bolt," for example, has 12 bolts on the cover, and only 10 on the differential.

General Motors installed a much stronger 10-bolt rear with an 8.5-inch ring gear on Buick turbo models and the Olds 4-4-2. These axles had a 30-spline design and were retained with a C-clip. Why these axle assemblies were not installed in the Monte SS is anybody's guess. In the Buicks, the gear ratio was 3.42:1, which is well suited for most applications.

The 4-4-2s usually have 3.73:1 gears. Some of these axles were limited slips while others were not. If you find one for sale, always properly identify it to be sure. The best way is to remove the rear cover, to verify if there is a Posi, and count the teeth on the ring and pinion. These rears are highly desirable because they can handle nearly as much power as an earlier 12-bolt rear.

This A-Body 8.5-inch rear axle assembly is installed in a 1971 Cutlass. The angle and placement of the upper control arm mounts is much different than on G-Body rears. The 10- and 12-bolt A-Body rear axle assemblies are often used in G-Bodies, but the housings are physically wider and the upper A-arms bind.

SouthSide Machine used to make an upper control arm that corrected this issue, but the company is no longer in business. I wouldn't use an A-Body 8.5 rear in a G-Body without the SouthSide Machine bars or something similar, due to the upper control arm angle differences.

Many question why General Motors didn't use the same bolt-in-style axle arrangement that's on early Buick, Oldsmobile, Pontiac (BOP) 8.5-inch axles, which were installed in 1971–1972 A-Bodies. C-clip eliminators are available for 7.5 axles. At higher power levels (typically more than 500 hp), many racers install a Ford 9-inch, which accepts a press-on axle bearing, and a set of aftermarket axles, to eliminate the problem of reliable axle retention when an axle breaks.

Swap and Upgrade Choices

G-Body owners often install earlier 10-bolt (8.2- and 8.5-inch ring gear) and 12-bolt (8.75-inch ring gear) rear ends from 1964–1972 A-Body models. But keep in mind that these axles are not an exact bolt-in and therefore require some modification to fit in a G-Body. These axle assemblies came in several widths; 1973–1977 GM cars with 8.5s are the widest and too wide for the G-Body. In addition, the cast-in "ears" for the upper control arms are not compatible with G-Bodies because the angle is different.

The four-link suspension used in the 1964–1972 A-Body and the 1978–1988 G-Body is dimensionally very similar. The lower control arm mounting points are identical, but the upper mounts are not, which results in severe bind when using most upper control arms.

To prevent coil binding, a kit was available (from the now-defunct SouthSide Machine Company) with "lift bars" for the lowers and an offset-designed upper control arm to accommodate this swap. They were often seen at drag strips around the country. This setup is fairly easy to duplicate, but it isn't seen much anymore because the earlier rear axle assemblies are less readily available for a reasonable price, and there are plenty of aftermarket bolt-in alternatives.

A GM 12-bolt rear is a popular choice for a modern rear end upgrade. Fortunately, several aftermarket suppliers, such as Strange Engineering and Moser Engineering, now offer these axle assemblies. The new housings have center sections with the correct upper control arm mount for a G-Body. In addition, they are made of higher quality iron than the originals for increased strength. The axle tubes are typically fully welded for added durability, and the axle shafts are the stock 30- or 33-spline versions.

The 12-bolt is the obvious choice to keep a G-Body an "all GM" car. It is relatively lightweight compared to many other aftermarket and OEM axles, particularly the 9-inch Ford and the Dana 60. It also is more efficient than a lot of other differential assemblies because the pinion gear rides lower on the ring gear, so there is less parasitic loss. However, it has a slightly weaker ring-and-pinion than a Ford 9-inch, so don't use a 12-bolt with C-clip axles in extremely high-powered cars, especially those equipped with manual transmissions and sticky tires. Although 12-bolt differentials came with C-clips, most aftermarket versions use a bolt-in axle.

Ford 9-Inch

Built from 1957 to 1984, the Ford 9-inch differential is the most popular rear axle assembly ever produced,

Note the shape and bolt pattern for this 12-bolt Chevy axle assembly with a Detroit Speed & Engineering cast-aluminum cover. For easy identification at a glance, look for two horizontal bolts at the top and bottom of the housing.

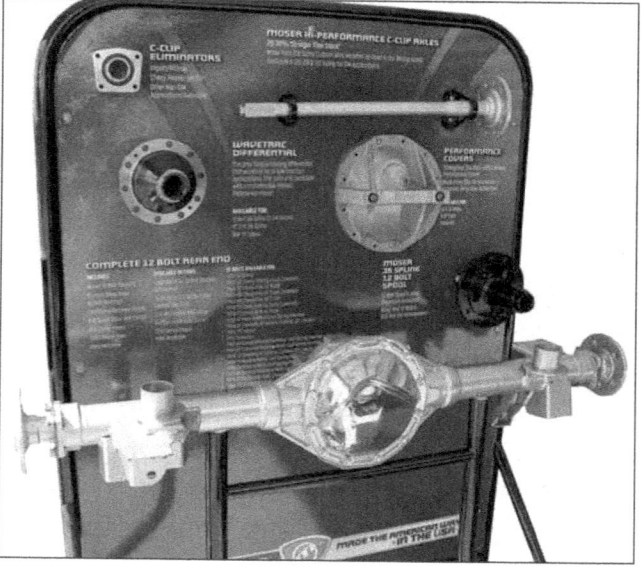

I prefer to order custom or fabricated rear axle assemblies from Moser Engineering, whether it is a Chevy 12-bolt or a Ford 9-inch axle. The 12-bolt shown fits a third- or fourth-generation F-Body car, but Moser also makes a G-Body housing. Moser offers custom widths and other options as well.

and it's been installed in many Ford high-performance passenger cars and light trucks. In fact, the performance differential aftermarket was built around it. Even though it is out of production, the aftermarket has stepped up and produced all the parts, including both stamped and fabricated housings. More gear ratio choices are offered for this axle assembly than any other in history, with new gears available in ratios from the mid 2.0s to 7.00:1.

Any differential type you can imagine is available for this application, including torque-sensing differentials, spools, and air lockers. The number of axle splines ranges from 28 to 40 (and bolt in, with very beefy axle bearings), with good differential choices up through 35. The 9-inch is relatively heavy and its pinion location is higher on the ring gear than a typical GM rear, so it is less efficient. However, this is also what makes it so strong.

I prefer this axle assembly for most applications when a non-OEM rear is installed. It provides incredible strength, but there are carriers that can be set up ahead of time and easily installed when needed. This is a big benefit for a car that's used for many applications because the owner does not have to deal with setting up clearances, pressing bearings on and off, and the other hassles involved in changing the gearset in a Salisbury-type rear.

Dana 60

The Dana 60 is another axle assembly to consider for extreme-duty drag applications. The very heavy Dana 60 has a 9¾-inch ring gear and is considered nearly bulletproof. Originally designed for trucks, it was used behind Hemi-powered Mopars in the 1960s and early 1970s. This axle is not common in Mopar passenger cars so it is not easy to find in salvage yards. This axle had limited its popularity to mostly drag racers, until narrowed rears became commonplace. Then people began narrowing truck housings for passenger car use.

Today, axle companies such as Strange offer new and improved versions of the Dana 60 called the S-60s. These axles are set up to bolt in to a variety of popular vehicles. In some cases, they are slightly less expensive than a Ford 9-inch because less beefing up is needed. Everything is heavy duty already. A comparably strong 9-inch Ford needs an aftermarket nodular housing, a Daytona-style pinion support, larger axles, etc., to equal the stock Dana.

Like the GM 10- and 12-bolts, the Dana 60 is a Salisbury-style axle assembly. It is rarely used for anything but drag racing or restoration.

Budget Option

Are there rear axle options for the budget-conscious G-Body enthusiast? Is there something for those who don't have a factory 8.5 or the $3,000 to do a good 12-bolt or 9-inch conversion?

Some have bought a 9-inch from a salvage yard and then modified it, but it's more difficult these days to find 9-inch axle assemblies in salvage yards. If you find one, you have to narrow it to fit the G-Body and weld on brackets to get it into the car. For a skilled fabricator, this conversion is not difficult, but it takes time and know-how. Once converted, you likely have a 28-spline 9-inch that needs a stronger case, limited-slip parts, new gears, and new bearings. This all adds up quickly, so it really isn't a "budget" option.

Late-model Mustangs have an effective differential setup and many G-Body owners are envious of it. Every V-8 Mustang from 1985 to the present came equipped with an 8.8-inch axle assembly, which are reliable and stout axles. Even Mustangs with turbochargers, superchargers, or nitrous don't seem to break them very often. The newer Mustangs come stock with 31-spline axles, disc brakes, factory Traction-Lok differentials, and decent performance gear ratios such as 3.55 and 3.73:1. It's too bad these rears don't fit a G-Body; or do they?

At first glance, the 1985–2004 style Mustang housings seem to be a viable "budget" option. They have a factory four-link suspension that is similar to the G-Body setup, and the width is close to the 8.8-inch axle. Unfortunately, the control arm mounting points are substantially different (especially the uppers), and attempts to use these housings (even with special, purpose-built control arms) have failed because of the lousy geometry caused by the misplaced uppers. The upper control arm mounting points are cast into the housing, and it isn't a good idea to weld new ones to a cast-iron housing. The 2005-present Mustangs use a three-link, and they are wider, so this isn't a good choice either. So, what to do? Well, there isn't an easy answer, but Bulldawg Musclecars is working on a solution. It recently constructed a jig to adapt an 8.8 Ford Explorer rear end to the 1964–1972 GM A-Body, and is currently working on a few changes to build a bolt-in 8.8 for the G-Body. The Explorer rear is commonly available

CHAPTER 8

in salvage yards, so it's readily available. It comes with 31-spline axles, large 3¼-inch axle tubes, Traction-Lok differentials, and 3.73 or 4.10:1 gears. Like the newer Mustang axles, it is equipped with disc brakes. The downsides are a Ford bolt pattern and C-clip axle retention.

The G-Body suspension geometry is retained. Any G-Body rear control arms work with this converted Ford 8.8-inch axle, so there is no need to change other components in your current setup. Housings are narrowed slightly to better fit the G-Body, and there's a choice of stock C-clip or bolt-in-style axles. Bolt-in axles have a GM Corvette-style brake option.

Stock Axle Assembly Upgrades

What if you just want to upgrade the rear end you have? The following are upgrades that apply to all the popular choices. I also include the reasons for each upgrade. In the case of the 7.5-inch axle assembly, I don't buy any of the listed upgrades because it's not worth the investment. Sure, you can purchase aftermarket axles, a good limited-slip unit, new gears and bearings, and a girdle-style cover. However, you still have a weak axle that any serious performance car will break with ease, particularly if you are using a manual transmission. I have broken them with less than 300 hp.

Moser Engineering built this Ford 9-inch rear for second-generation F-Body cars, but it is also available in a G-Body configuration that is a direct bolt-in. This one is equipped with a back brace for extra strength plus drain and fill plugs, which are all extra-cost options. The back brace isn't needed for most applications, but the drain and fill plugs are a must.

General Motors equipped G-Body cars with either the weak 7.5 or scarce 8.5 axle assemblies. The stock 8.5 assemblies are rather expensive and aftermarket axles come at a high price. Bulldawg Musclecars offers a high-performance, affordable alternative, the Ford 8.8 (shown). These are fully jig welded for accuracy and to preserve the factory geometry.

A fabricated aftermarket 9-inch Ford is a popular choice for drag cars because it delivers the necessary strength to cope with high-horsepower and hard drag-strip launches. Chris Alston ChassisWorks builds direct bolt-in housings for a factory rear replacement or setup with a drag-style parallel four-link.

You will also spend at least $1,200, not including any labor charges for setup, and you'll eventually buy one of the stronger rears to replace it. Take my advice: Save your money and apply it to a rear with some potential. You can probably sell the stock rear to someone at the drag strip who doesn't heed my advice, and is putting together yet another 7.5-inch.

Ring-and-Pinion Gears

Many replace the ring-and-pinion gears with new ones to arrive at a desired ratio. The vast majority of G-Bodies came with dismal gear ratios, but performance models were equipped with effective gear ratios, and some were as low as 3.73:1 from the factory.

Select a gear ratio that matches the car's intended use and the type of transmission you plan to use. Any of the overdrive automatics (TH2004R, TH7004R, or 4L60/65/70/80) have either a .67 (TH2004R) or .70:1 (all others listed) fourth gear. This allows a lower (numerically higher) rear end gear ratio without any ill effects from a conventional 3-speed automatic, such as the TH350 or TH400, which are 1:1 in third gear. Don't go too far, though, as the first-gear ratios are lower than in 3-speed automatics, and the extra torque multiplication results in excessive wheelspin.

In most G-Body applications with common GM automatic transmissions use 3.42 or 3.73:1 gears. A manual transmission, such as a T-56 6-speed, can have a sixth-gear ratio as high as .50:1, so it can use a much lower (numerically higher) gear than even the overdrive automatics. 4.10:1 gears are typically used in these applications because highway mileage doesn't suffer and acceleration is greatly improved.

Driveshaft speed increases exponentially with lower gears, so don't select a gear ratio that's excessively low (numerically high). They put more stress on the drive axle, and this results in vibration and potential catastrophic failure at high speeds, which negates any perceived benefit. I once tried 4.56 gears in an F-Body with a T-56. It was equipped with a very lightweight and well-balanced aluminum driveshaft. Although acceleration was awesome at low speeds, the car shook violently between 85 and 110 mph.

Pinion Supports and Yokes

The stock Daytona-style 9-inch pinion support is fairly beefy, but as a relatively inexpensive upgrade, it is worth doing. It uses larger pinion bearings than the stock pinion support. Increased gear life, reduced friction, and overall durability are all results of an improved pinion support.

One upgrade that is often overlooked is to the pinion yoke, even though it's an important component. Most axle assemblies need a bigger, stronger U-joint for a higher-horsepower car, and it quickly becomes the weak link in the drivetrain.

Most upgraded pinion yokes use the 1350-series Spicer joint, which is somewhat the standard among drivetrain parts suppliers. It is likely what any high-performance, custom-built driveshaft comes with. I consider it a must for any high-powered car (400 hp or more), particularly one with a manual transmission. Be sure to run a transmission yoke set up for the 1350-series joint, so that the front joint doesn't become the weak link in the drivetrain.

Axle Shafts

Installing a set of stronger axle shafts is another popular and often required rear end modification. They are typically made of higher-grade

Aluminum Daytona-style pinion supports are available for smaller spline count axles, not just 35 as shown here. I always add this when building a Ford 9-inch because it is a fairly inexpensive upgrade. I usually go overboard on the drivetrain components because my vehicles often wind up producing more power than originally anticipated. It is cheaper to install these components the first time, rather than going back and replacing parts later.

Most people don't think much about the driveshaft, but if you want to use a stronger 1350-series universal joint, you need to change the yoke. This is good, cheap insurance.

steel than stock axle shafts and feature higher spline counts. More splines equal increased strength, and the larger axles usually cost about the same as smaller ones.

The differential also needs to be upgraded to the higher spline count. In some cases, axles are available in a choice of retention methods: C-clip or bearing. Again, cost isn't a significant factor, but correctly installing bearing-style ends in the housing is beyond the ability of the average enthusiast.

As a general rule, when building a new rear, select the largest axles you can, with the limiting factor being the differential type that you want to run. For example, 31-, 35-, or 40-spline axle shafts are available for most Ford 9-inch axles housings, but the higher spline counts have far fewer choices in differentials.

This probably won't be an issue if you are running a spool, but for anyone else, it's a major consideration. In some cases, a larger case (9-inch in particular) is required.

Wheel Studs

Adding stronger wheel studs is an important upgrade to consider, especially when buying new axles. I typically order half-inch studs whenever a new rear or a set of axles is ordered, for the extra peace of mind they afford. Longer studs, required in many racing classes, are a good idea as well.

Differentials by definition serve to allow different wheel speeds for each axle. Essentially, they engage and disengage as needed to allow the vehicle to turn smoothly because the inside wheel doesn't turn as many revolutions as the outside wheel. Non-locking (or open) differentials are the most common type found in G-Bodies and the majority of non-performance passenger cars. But these differentials keep applying power to the wheel that is spinning. In other words, it's the wheel that does the one-tire burnout. Therefore, these are not the choice for high-performance applications.

Limited-Slips

A limited-slip, or Posi-Traction, differential is high on the upgrade list for any high-performance axle assembly. What type is best for you depends on your intended usage, driving style, and power level. Limited-slips allow for excellent traction, without the harshness and abrupt locking and unlocking found in a mechanical-type locker. They are best suited for daily driving, autocrossing, most forms of road racing, and light drag racing. Several manufacturers, using a variety of trade names, make a limited-slip differential.

Limited-slip differentials fall into one of two categories according to their type of friction material: clutch and cone. The clutch type is the most common, and it uses a stack of replaceable clutch discs. Most GM and Ford OEM units used this type of limited-slip, and they work well. One of the major benefits is that they are easy and inexpensive to rebuild.

The cone design is most often seen in Chrysler products in OEM applications. These units also work very well, but aren't serviceable in the field as a clutch type is. Auburn Gear offers rebuild services, even though they rarely need rebuilding.

I have used both types with great success, but if you plan on lots of drag racing I lean toward the clutch type, so you can rebuild it yourself if needed.

Detroit Lockers

When discussing locking differentials, the famous Detroit Locker is usually the first that comes to mind.

An Eaton clutch-type limited-slip differential is by far the most common type used, and has the advantage of being rebuildable. It is a suitable unit for most street- or track-driven cars, and can take lots of abuse.

The legendary Eaton Detroit Locker is known for its tremendous strength, but is twitchy on the street as the gears lock and unlock. Some suppliers offer a "soft" version, but in my experience they still provide a hard gear lock. I recommend the soft versions of the Detroit Locker only for radical street/strip applications.

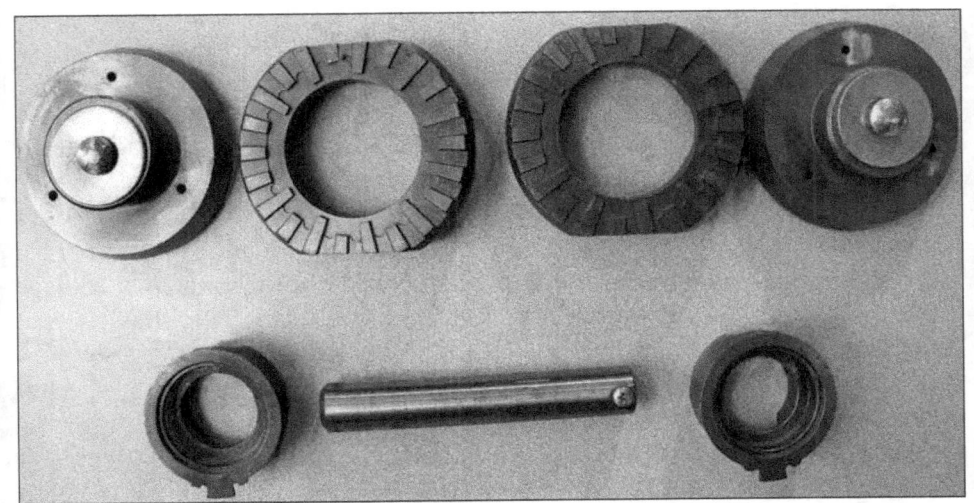

Similar in operation to the Detroit Locker, the Powertrax locker fits in a stock open-differential case and replaces the spider gears. For this reason, it is very easy to install, and doesn't affect gear setup in any way. Unfortunately, the unpredictable handling characteristics and noise aren't worth the simplicity of installation.

If you're building a G-Body for drag racing, a full spool is a viable option. The lack of differentiation leads to instability in turns and can lead to loss of control. They are also very hard on axles, for the same reason.

Lockers work great in extreme-duty applications, such as off-road, drag racing, and in some cases, road racing. While the designs differ from one manufacturer to another, they all use interlocking gears to mechanically lock the axles together, and therefore the differential sends equal power to each wheel.

Detroit Lockers are strictly mechanical, and it's sometimes a very noisy, harsh process. They use a central spider gear, with an interlocking clutch (basically, another gear that meshes with the spider gear) on each side. They are spring loaded and stay locked as long as the vehicle is going straight, but the outer wheel on a turn unlocks. ARB lockers use air pressure, or even electromagnets, that allow the driver to choose when to lock or unlock the differential.

This technology is most common in off-road applications, but can work as well in a dual-purpose street car. These units feature a positive engagement rather than a wear material engagement, so there is little wear.

I can't stress enough the harshness of these units. The noise and sometimes unexpected locking and unlocking of the axles can be a bit unnerving. I only use this type of unit in the most serious of street cars.

Spools

Technically, a spool isn't a differential at all. A spool is a machined steel collar that not only has splines to accept the axles but also provides the mounting surface for the ring gear. The axles are permanently locked together, and this arrangement provides no differentiation at all. The only time a spool is desirable is in drag racing or dirt track racing. This is *not* a street-oriented piece and could lead to broken axles if used much for street driving.

A mini-spool is a similar device, and shouldn't even be used for drag racing. Like the spool, it positively locks the axles together, but it doesn't replace the entire differential. Instead, it replaces the spider gears of an open rear. By necessity, these

The mini-spool is designed for use in dirt-track applications and should never be used in any other application, including on the street. They break under hard usage on pavement.

units are very small and very weak. Never use them on a paved surface.

Differential Covers

A stamped-steel rear cover is typically fitted to Salisbury-type rears, and it serves only to retain the fluid and keep dirt out of the gearset. It does not add strength. Because the cover has to be removed to service the gears, running a back brace for

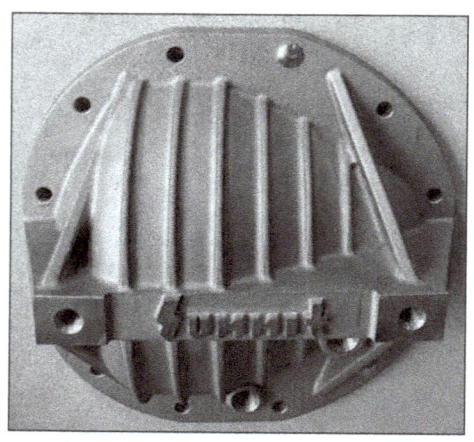

This cast-aluminum cover accepts set screws that can be pre-loaded against the main caps, preventing movement under high-torque loads (and the resulting breakage). Without these preload caps, a cast-aluminum cover offers little real benefit.

added strength (as often seen on 9-inch rears) isn't possible. There is a solution, but like just about every other performance part, the various types aren't created equal.

A thicker, cast-aluminum cover is a popular upgrade for Salisbury types of axle assemblies. Some are finned and tout a cooling benefit, but any gain here is marginal. Set screws that can be pre-loaded against the rear's main bearing cap come with the aluminum covers. This prevents the natural tendency for the set screws to "walk" as the pinion tries to climb the ring gear on hard acceleration. The better ones come with a stud kit that replaces the main caps' stock bolts, especially handy if the unit is to be disassembled often (such as in a drag car).

A few covers have cast-in bosses, and a heim-jointed rod extends to a supplied tab, which is welded to the axle tube. This gives some of the rigidity of a back brace and adjustability. These systems are available from LPW and others for a variety of differentials, but they are most often seen on the 8.8-inch rears of high-powered Mustangs. This is a worthwhile upgrade to any Salisbury-style rear that sees hard usage.

Back Braces

Hotchkis-style differentials, especially those with stock housings, can suffer from flex. This results in broken gears, limited-slip differentials, and axles. A back brace, at least in theory, is the easier solution. Typically, a back brace is constructed of a U-shape bar of steel tubing, bent to connect the outer part of the axle tube to the rear center.

Typically back braces are cut so they fit very tightly to the housing and axle tubes, and can be welded along the entire perimeter. They essentially reinforce both axle tubes, and tie them into the center section. This adds a great deal of strength and prevents housing flex.

But take care during the installation process; otherwise, the heat from welding on the brace invariably warps the housing. The careful application of pressure from a hydraulic press can easily correct this, if you know what you are doing.

Many companies, such as Moser Engineering, add a brace to any new 9-inch housing, and it is worth the extra expense to have professionals install it if you're not well versed in the procedure.

Gear Cases

An upgrade that applies to 9-inch differentials is to the gear case itself. The "nodular" case is the most desirable, and it's found in the highest horsepower muscle cars and heavy-duty trucks. I suppose there is a chance of finding one in a salvage yard, but I personally have never found one there.

Moser, Strange, Currie, and many others, offer their own version of the nodular iron case. These aftermarket cases use higher quality materials than the originals, and design improvements add strength where needed.

Install a high-performance axle assembly if your G-Body is going to produce more than 400 hp. However, if the car produces less than 400 hp, you can get reliable performance out of an OEM rear differential.

Buying an aluminum case is a viable but more expensive option. They are mostly installed on show cars, or on drag cars or road racers in which reducing unsprung weight savings is of utmost importance. While strong, they do flex more than a comparable nodular case. This deflection throws the gears out of alignment, creates more wear, and leads to reduced gear life. While this may not be a concern in a rebuilt differential in a high-horsepower drag car, this frequent maintenance is not desirable for an unsponsored street car or bracket racer.

If you have a Hotchkis-style rear and need a stronger housing, install a back brace. It's best to have a professional do this because the welding required for installation can easily warp the housing, requiring it to be straightened. Moser can add this as an option to any of its rears.

From left to right: a stock center section, a through-bolt aluminum center, and a nodular iron unit. The nodular iron 9-inch differentials are strong enough for 400 hp and much more. Aluminum centers have improved greatly, but they do tend to flex a bit in high-horsepower applications, and this creates more gear wear and a shorter life span. Moser builds all of these center sections.

The Ford 9-inch axle's drop-out center section makes it one of the most popular axles because you can swap center sections with relative ease. Owners prepare different center sections at home, then at the track they don't have to do any setup at the time of the swap. This comes in very handy for a street/strip car because the center can be changed out in 30 minutes (or less).

Moser Engineering built this particular unit, which features a nodular iron case, Daytona-style aluminum pinion support, Auburn Pro Series limited-slip differential, and 3.75:1 gears. The pinion yoke is the larger 1350 style.

If you are having a rear built, the preferable unit to use is a nodular iron case by Moser Engineering.

If you are concerned more with weight than parts longevity, or are building a show car (they look very nice polished), this Moser Engineering aluminum case is a suitable unit.

CHAPTER 9

GET INTO GEAR
MANUAL AND AUTOMATIC TRANSMISSIONS

The vast majority of G-Bodies came with automatic transmissions. The early ones have lockup versions of the TH200 and TH350 (both 3-speeds) while the later ones mostly have versions of the TH2004R. Because of the fragile nature of the earlier transmissions, most have long since been replaced with a stronger and more reliable transmission. Never take for granted that your car has a certain transmission, without checking first.

The Powerglide (2-speed), Turbo-Hydramatic 350 (3-speed), and TurboHydramatic 400 (3-speed) are the most common, non-overdrive automatic transmissions in use today.

Powerglide

In stock form, the Powerglide is suitable for high-performance builds and should only be considered for a restoration. With only two speeds, the engine is rarely in its powerband with any streetable differential gears. They are very popular with drag racers because they transmit a lot of power and have a huge aftermarket. Most importantly, the gearing doesn't shock the tires as badly as a 3-speed does in a high-horsepower drag car. They work best in lighter cars, with very low gears, or with large turbos or superchargers. They have no place in a street-driven car.

Turbo Hydra-Matic

Since the mid 1960s, the Turbo-Hydramatic series of transmissions have been used in GM vehicles. The TH350 and TH400 covered here are 3-speeds, with no overdrive. The rest are 4-speed overdrive transmissions.

Turbo Hydra-Matic 350

The TH350 is similar in size to the Powerglide; both share approximately the same overall length, case size, and transmission mount location. The relatively lightweight and efficient TH350 is by far the most popular 3-speed automatic ever offered by General Motors. It is

BorgWarner originally made the T-56. Tremec now makes it. Even in stock form, it is very strong, and aftermarket versions can hold 750 ft-lbs of torque or more. This one is going into the *Sabre*.

GET INTO GEAR

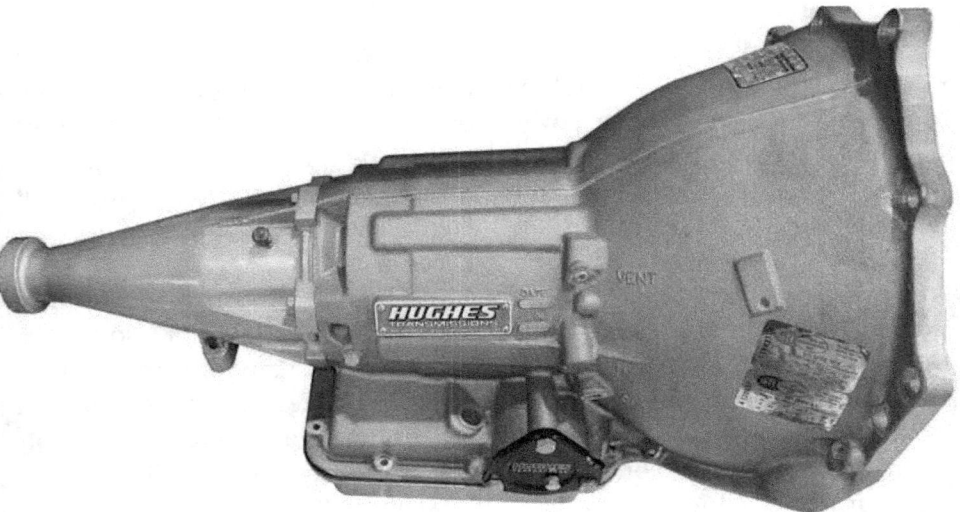

This aftermarket Powerglide has the same dimensions and overall appearance as an OEM unit. I would only use this in a drag or circle track car because this 2-speed automatic doesn't have a variety of gear ratios required for high-performance street driving, autocrossing, or road racing.

available in two different lengths: a "short-shaft" version for passenger cars and a "long-shaft" version for trucks.

For G-Bodies, stick with the short shaft. Like the Powerglide, it has huge aftermarket support and can be beefed up to handle considerably higher power levels than it could in stock form. The vast majority of G-Bodies on the road use this transmission. It is far cheaper to buy and to rebuild than any of the overdrive transmissions. It easily replaces any of the other automatics the cars came with.

The TH350 comes in both Chevrolet and BOP (Buick, Oldsmobile, Pontiac) bellhousing patterns, so it's compatible with any G-Body factory-installed engine family. Some, but not all, are drilled with bolt patterns. In any case, make sure that the bolt-pattern of the transmission you are installing matches the bellhousing. Keep in mind, too, that adapter plates are available and fairly inexpensive (around $80). So if you have a good transmission with the wrong pattern, it can still be used.

The only real downside to this transmission, other than the lack of

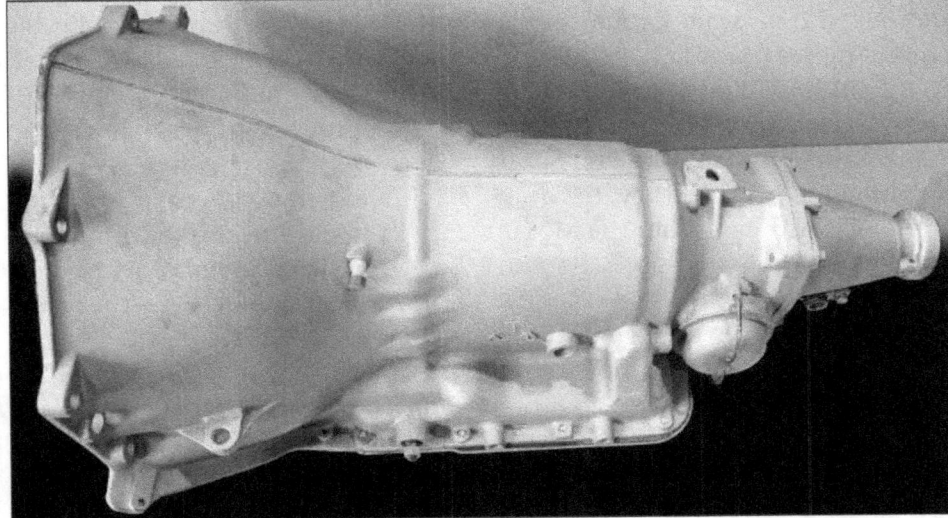

The Turbo Hydra-Matic 350 has nearly the same overall length as the Powerglide and the overdrive TH2004R, making swaps easier. This is the most popular non-overdrive automatic because it is cheap to build and durable.

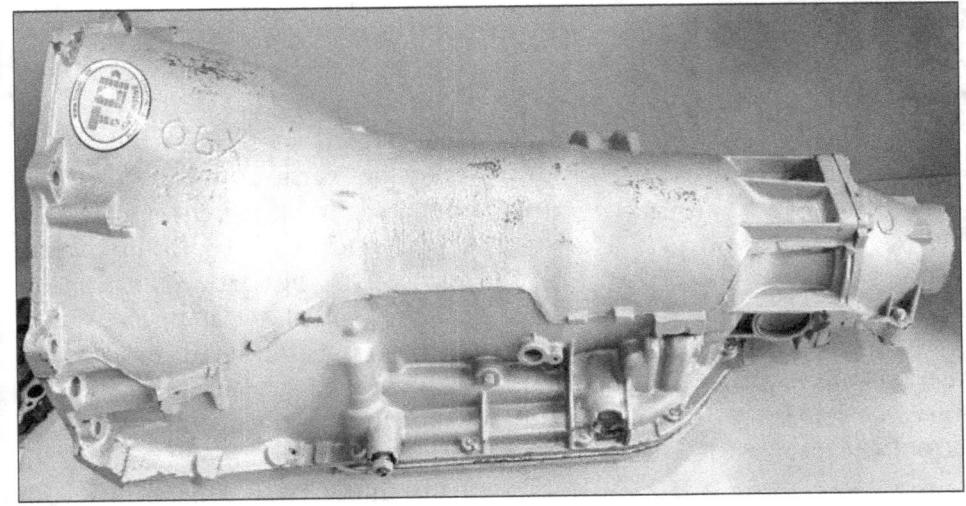

The Turbo Hydra-Matic 400 has a case that's much larger than that of the TH350. It was designed to go behind big-block engines, and while very durable, it robs a lot of power. It is best for big-block swaps or high-horsepower drag applications where a third gear is desirable.

overdrive, is that it wasn't designed for more extreme, high-torque applications. It can be built to handle even the nastiest of big-blocks, but this gets expensive fast. For a typical performance-built TH350, you're good for about 400 hp. More than that and you need something stronger, such as the TH400.

Turbo Hydra-Matic 400

The TH400 is a good choice for extreme applications. It was designed for big-block applications, although small-blocks can use them just as easily, and heavy-duty vehicles, such as trucks and RVs, use them. The TH400 also comes in long- and short-shaft versions, and BOP and Chevrolet pattern versions, similar to the TH350. The case is much beefier and still clears G-Body tunnels with no issue.

To fit a TH400, it is necessary to move the transmission crossmember back by simply shortening the driveshaft and adding a larger yoke. A conversion U-joint easily mounts to the larger yoke, so a different driveshaft isn't needed. The TH400 is heavier and absorbs more power than the TH350, but it handles more power for less cost than a comparably strengthened TH350.

Turbo Hydra-Matic 2004R

From the mid 1980s, the TH2004R was found in many G-Bodies, including the Buick Grand National. It is a 4-speed overdrive automatic, with very favorable gear ratios for performance (2.74, 1.57, 1.0, .67:1). These transmissions operate well with most performance gear ratios (most commonly 3.42 and 3.73:1 in the G-Body), and they are pretty strong from the factory. The Buick turbo versions are the most sought after because they have an improved valve body, but the Monte Carlo SS and 4-4-2 versions are also very good.

The differences are in the fluid passages themselves, so they aren't things that can be overcome with a "shift kit." Although Chevrolet-only patterns have been rumored, I have never actually seen one, and all the ones that I've seen have a dual-bolt pattern (Chevrolet and BOP). This makes it great for use behind BOP engines because no modifications are necessary.

This transmission is roughly the same length as the TH350, so a swap doesn't require any driveshaft modifications (yokes are the same), but the mount is located farther back on the tail housing, as on a TH400.

The TH2004R uses a throttle valve (TV) cable to control line pressure through the gears. It looks like a kickdown cable, but it is vital to the proper operation of the transmission. Improperly adjusted, it leads to a burned-up transmission.

There's an ample supply of aftermarket parts for this transmission, but it isn't as popular with most transmission builders as the more common TH700R4.

Turbo Hydra-Matic 700R4

The TH700R4 (later known as the 4L60) was never factory installed in the G-Body, but I cover it here because it is undoubtedly the most popular, non-computer controlled overdrive transmission for retrofit into any GM vehicle.

The first-gear ratio is very low, at 3.06, with a 1.62:1 second, 1:1 third, and .70:1 fourth. Like the TH2004R, it uses a TV cable. Early 27-spline versions were weak, but the later 30-spline versions are very strong, and can be beefed up considerably with heavy-duty sun shells, clutches and steels, and billet shafts. This transmission's parts interchange with the later electronic versions, which include the 4L60E, 4L65E, and 4L70E. These are progressively stronger than the TH700R4.

The TH700R4 was never factory installed in the G-Body. It's the most popular of the non-electronic GM overdrive automatics; one reason is that General Motors has continually refined it over the years. You can install many of the parts for later 4L60/65/70E transmissions in the TH700R4. A throttle-valve cable determines the line pressure.

The new Supermatic 4L70E is dimensionally similar to the TH700R4, but a separate bellhousing, similar to most 4L60E and 4L65E transmissions, differentiates it from the TH700R4. In addition, it has no throttle-valve cable or electronic control.

To get the most out of an automatic, the torque convertor stall speed should match the powerband of the engine, which is largely determined by the camshaft. If the stall speed and powerband are not matched, accelerating from a dead stop is difficult. It stalls often, is virtually undrivable at low speed, or at the very least is much slower.

The bellhousing patterns are Chevrolet only, but similar to the other Turbo Hydra-Matic transmissions, so an adapter plate can be used if needed to bolt it up to a BOP engine.

They are longer than a TH350 or TH2004R, so the driveshaft needs to be shortened, but the mount location is similar to that of a TH400.

The stock crossmember is easily moved back, or you can use an aftermarket crossmember, such as those sold by G-Force Racing Gear and others.

Later transmissions used a vehicle speed sensor (VSS), rather than a speedometer cable, but you can retrofit one to the other if needed.

New-Generation Transmissions

The aforementioned 4L60E, 4L65E, and 4L70E are dimensionally similar to the TH700R4, but all use a removable bellhousing, except the earliest versions. They have an ECM to control line pressure rather than a TV cable. You need an add-on controller that costs from $650 to $1,000 for a carbureted application or any car that doesn't have an ECM capable of supporting an electronic transmission. For this reason, they are most often seen behind LT1 or LS-series swaps, because these engines were factory equipped with them. Mounting and bellhousing patterns are the same as on the TH700R4.

4L80E/4L85E 4-Speed

The 4L80E/4L85E is GM's strongest electronic 4-speed overdrive transmission, and it is based on the

Match your vehicle specs to a particular transmission, which requires coordination with the transmission supplier. Work closely with the supplier, providing your vehicle specs, such as gearing, weight, cam specs, intended usage, etc., so that you use all the available power and maintain good street manners.

This TCI unit is very streetable as long as everything else is a good match, though I tend to stay with lower stall speeds for street-driven cars. Use a good auxiliary transmission cooler when running a high-stall convertor.

A deeper transmission pan, with a larger fluid capacity, also helps to keep the transmission running cooler. Heat kills transmissions, and anything that keeps the temperature down is a good idea, especially with a high-stall convertor.

CHAPTER 9

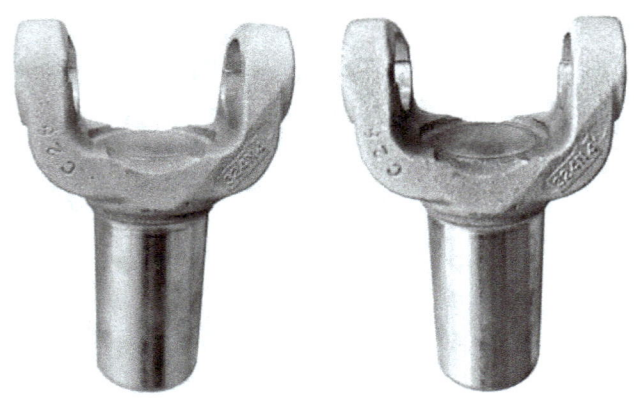

When swapping to a TH400 or 4L80E/6L80E, a larger yoke is needed. This beefy unit from Moser also allows the use of 1350-series U-joints for extra strength.

Running an overdrive unit with a lockup convertor? This B&M unit allows the convertor to be manually locked or unlocked by the driver. This is a great way to retain the advantage of a lockup convertor (lower engine cruising speeds and better mileage) without having to deal with it in competition, where the locking and unlocking may not be wanted. (Photo Courtesy Doug Lutes)

TH400. Trucks, big-blocks, and 6.0-liter LS-series vehicles were mated to this transmission. The gear ratios for the first three gears are identical to the TH400, and it has a .75:1 overdrive. An ECM or a stand-alone controller controls it.

The 4L80E withstands massive amounts of torque, but it comes with a tradeoff: The transmission is very large and heavy. Some clearancing is necessary to install it on a G-Body. That said, it is still the best choice for any high-powered street car, at least from a durability standpoint.

6L80E 6-Speed

The 6L80E is a 6-speed automatic with many of the same features as the 4L80E, and two additional gears. This transmission will undoubtedly become very popular as time goes on, but at this time isn't very popular for swaps. This is mainly due to the cost, increased size and weight, and complexity. I have yet to swap one of these transmissions, so I can't comment further.

OEM Manual Transmissions

Some of the earlier G-Bodies came from the factory with 3- or 4-speed Saginaw manual transmissions with floor-mounted shifters. These used conventional mechanical clutch linkage.

These two Saginaw transmissions are not suitable for high-performance applications because of the light-duty design and gears, including a very low first gear and wide ratio splits. They came with typical factory shifters, with a very loose feel. They can be easily improved with an after-market shifter, such as a Hurst.

These trannies are adequate for daily driving or cruising, but anything more than that causes durability issues.

BorgWarner

The BorgWarner T-5 was the first prolific and popular OEM 5-speed manual overdrive transmission. In GM applications the T-5 uses an internal rail shifter (similar to transmissions of today) and hydraulic linkage. Fords used the T-5, but they were mostly cable actuated. The T-5 was used in third-generation F-Bodies, but interestingly enough, it was not used for any 5.7-liter (350) applications.

General Motors only installed the T-5 in the 305 and V-6 cars because it couldn't handle the added torque from other high-performance engines. I don't recommend the T-5 unless you already have all the components needed, and don't plan to make much power. The Ford versions are a bit stronger, especially in later models. While they can be adapted, the cost of the components adds up quickly, so there are better choices for the money.

In 1993, General Motors installed the BorgWarner T-56 6-speed transmission in the F-Body. The 1993 models had a .67:1 sixth-gear ratio, while all later OEM versions used a .50:1 ratio. Like the T-5, they came with hydraulic clutch actuation, which is very easy to retrofit, and a 26-spline input shaft but use the smaller output shaft so a driveshaft from an automatic car (non-TH400) works. Length is the same as a TH700R4.

This transmission is extremely strong, even in stock form, and aftermarket fixes are available for the problem areas. Also, the .50:1 sixth gear allows much lower rear-end gears than in many other transmissions. I prefer 4.10/4.11:1 gears with this transmission, yet there's no compromise in cruising RPM or MPG. Cruising speeds at 85 mph of around 2,300 rpm are common, and MPG is in the mid to high 20s, even with a large cam and head work.

Be aware that all T-56 transmissions aren't the same; there are 1993–1997 LT1 versions, Dodge Viper versions (both OEM, and aftermarket versions that fit other makes), and several LS-series versions. Be certain of what you are buying and what fits best for your application. (See Chapter 10 for more swapping details.)

Muncie

A Muncie 4-speed is a direct bolt-in replacement for a Saginaw transmission in a G-Body. Most Muncies use the same 10-spline clutch disc as the Saginaw, and there are three basic versions available: the M-20 (wide ratio), the M-21 (close ratio), and the M-22 (close ratio, with heavy-duty straight-cut gears). The M-22, known as the Rock Crusher because of its distinctive high-pitched whine, is the strongest, but the M-21 is preferred in most cases since it isn't as noisy and is nearly as strong.

Although these transmissions are no longer produced, they are easy to find on the used market and most of the parts are available new. Many are subjected to extreme use, so if you want a Muncie the best source is a rebuilder who specializes in them.

The BorgWarner Super T-10 was GM's replacement for the Muncie, and personally, I prefer it to the Muncie. The Super T-10 uses a 26-spline input shaft, which is stronger than the Muncie's input shaft, and it uses the same heavy-duty yoke as a TH400. So, to replace one of the Muncie or Saginaw 4-speeds, you need a new clutch disc, a larger yoke, and a conversion U-joint. Depending on the length of the differential yoke, you may also need to shorten the driveshaft slightly, but it's not necessary in most cases.

The Super T-10 and the Muncie M-series transmissions are somewhat difficult to find in used OEM form because they were installed in select GM high-performance cars only, such as late second-generation and very early third-generation F-Bodies and early 1980s Corvettes. A 4-speed transmission was a fairly rare option in these cars. Fortunately, Richmond Gear still manufactures a transmission for it, so you can buy a new one.

Aftermarket Manual Transmissions

New aftermarket transmissions are often much stronger than OEM transmissions and offer a variety of gear ratios for specific high-performance applications. In addition, many new manual transmissions offer a fifth or sixth gear for greater fuel efficiency.

Richmond Gear

The Doug Nash 5-speed was the first commonly available, heavy-duty 5-speed manual transmission for GM passenger cars. Today, Richmond Gear builds and sells this transmission as the 4+1 Street 5-speed, and it was considered the strongest manual transmissions until the BorgWarner T-56 (now produced by Tremec) was released. Although it does have five forward speeds, it isn't an overdrive. Rather, the first-gear ratios range from 3.27 to 4.41:1, negating the need for a low rear end gear. The fifth gear ratio is 1:1. This transmission is the same length as a traditional 4-speed Muncie or Super T-10, and it has 26 splines similar to the Super T-10.

Mechanically, the Richmond installs in a G-Body similar to a 4-speed, but you may need to modify the tunnel (basically split it down the middle, and weld in a filler piece) to accommodate the larger case.

Richmond also builds a 6-speed version of this transmission, with a higher first gear than the 5-speed, and with sixth gear being an overdrive. Both use external shifter linkage.

Of the non-overdrive manual transmissions, the Muncie 4-speed is the most popular, mainly due to its use in various muscle cars of the 1960s and early 1970s. This unit is from a 1969 Z/28. It is a direct bolt-in for a manual-transmission G-Body.

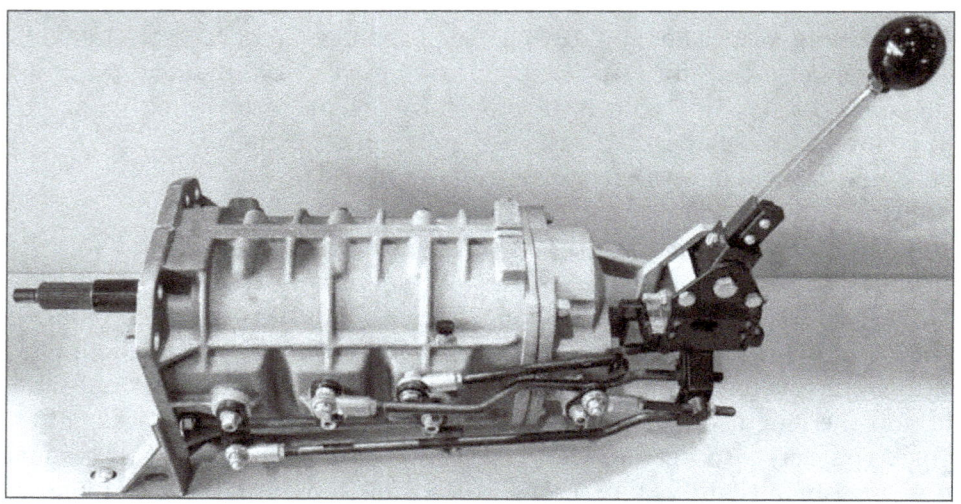

This Richmond 6-speed is a good transmission, but only has one overdrive. It is based on the old Doug Nash 5-speed, which has no overdrive. It can use a stock 4–speed crossmember, or a new one can be fabricated if your car has an automatic.

Tremec

The Tremec 5-speed is one of the most popular manual transmission choices for any make or model. Over the years, there have been many different versions, but today they are known as the TKO-500 and TKO-600. The number refers to the torque rating, and there are different gear ratio choices for each.

Like the T-5 and T-56, the 5-speed uses an internal rail shifter. This transmission is a very good replacement for a factory 4-speed, as it usually requires little, if any, sheet-metal alterations to install it. Aftermarket offset shifters are available to put the handle exactly where you want it. The transmission even has additional mounting points for the shifter itself. Like most other heavy-duty transmissions of the past thirty-plus years, it uses a 26-spline input shaft. It's a bit longer than a traditional 4-speed, so driveshaft shortening is needed.

The Tremec 5-speed was originally designed as a stronger replacement for the T-5 in Ford applications, but soon became popular as a retrofit into GM and Mopar vehicles as well. The Ford Racing Tremec was originally spec'd for the Cobra R. They were popular for swaps at one time, and there are still many of them around, but now GM-specific versions are available for easier installation.

Tremec 5-speeds, such as this American Powertrain unit, are now offered in two basic versions: the TKO-500 and the TKO-600. The difference is in torque ratings as specified by the number in the name. And the gear ratios are different. Some choice is available, but the TKO-500 always has a 3.27:1 first-gear ratio, which makes it work best in cars with numerically low axle gear ratios.

CHAPTER 10

A HIGH-PERFORMANCE TRANS-FORMATION
MANUAL TRANSMISSION SWAPS

As great as the newer automatics are, there is just something about a manual transmission in a hot rod, and there is nothing better for a car that is going to be used for autocrossing or road racing. Unfortunately, not many G-Bodies came with them, and finding a good set of pedals and linkages in the salvage yard is highly unlikely.

Fortunately, you can install a manual transmission without any factory parts. Several sources offer complete kits, including the pedals that duplicate the original mechanical clutch linkage. If you are planning to use a small- or big-block Chevrolet, it's a great way to go. Rick Bejarano of Metco Motorsports used original-style parts when building his 502/Tremec 5-speed Monte Carlo SS. If you are using a 4-speed manual transmission, a 5-speed Tremec, or one of the Richmond 5- or 6-speeds, the mechanical linkage eliminates the headache of building or modifying a hydraulic system.

G-Force and others build crossmembers, or you can build or modify your own. Most pre-made ones maintain a proper driveline angle when installed, so this is a little less of a concern, but *always* check it. The actual angle is important but it's absolutely critical to keep the

Holley Performance Products owns this El Camino that features an LS3 engine with a T-56 transmission. The swap was completed using mostly parts available from Holley, including engine mounts, headers (Hooker), and accessory brackets. For me, a manual shifter and a third pedal are always welcomed sights in a G-Body.

transmission and rear end yokes parallel to one another. A 3- to 5-percent angle usually works well.

A universal crossmember, with shims as needed on the transmission mount, can resolve the problem and maintain transmission and rear end alignment, saving several hours of fabrication. You just want to watch for exhaust clearance, and check the driveline angle before finalizing anything.

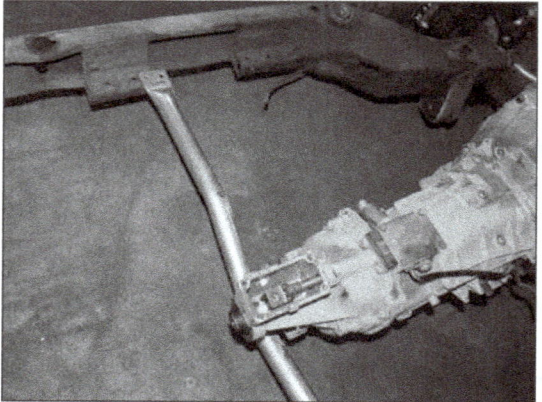
This lime green T-56 came from a 1998–2002 F-Body, but it found a new home in a G-Body. The BRP/Musclerods LS engine swap kit includes a transmission crossmember, which is available for all popular transmissions. G-Body frame hardware changed over the years; early models, such as this 1983, had two different crossmembers. BRP makes both of them. Note that the one installed here is wrong for this chassis. However, in its final form this car is going to have the engine set back considerably (for turbo plumbing clearance) so the crossmember won't be needed. If you box your frame, be sure to leave room to get the crossmember in and out for service.

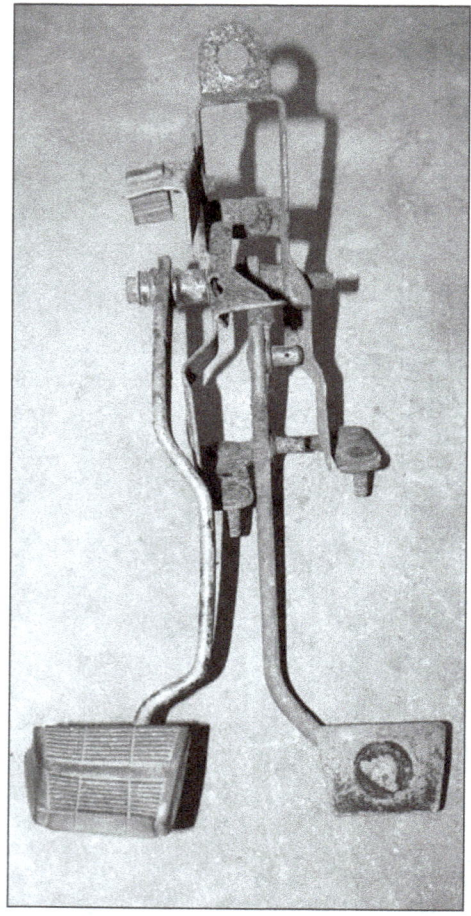

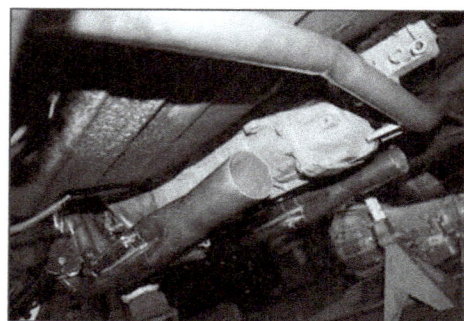
There is plenty of room for the exhaust to pass under the crossmember, even with a 3-inch exhaust.

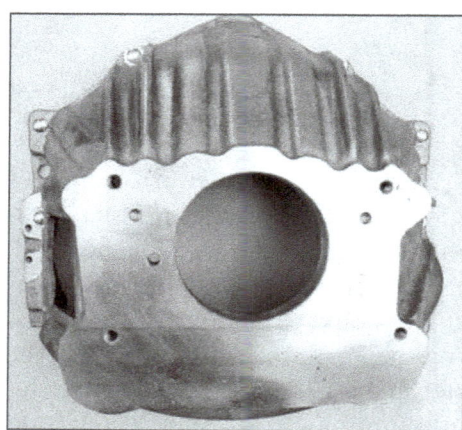
This Tremec cast-aluminum bellhousing is typical for a stock transmission. While it serves the purpose of connecting the engine to the transmission, I prefer not to use this type of unit, for safety reasons. If there is a clutch explosion, the aluminum will not stop it.

Though this set of factory pedals is a bit crusty, it can be restored and look as good as new with a little time in the bead-blast cabinet and some paint. The pedal bushings are in good shape, but new ones are available from the restoration aftermarket if needed. Reproductions of these pedals are available, usually as part of a kit.

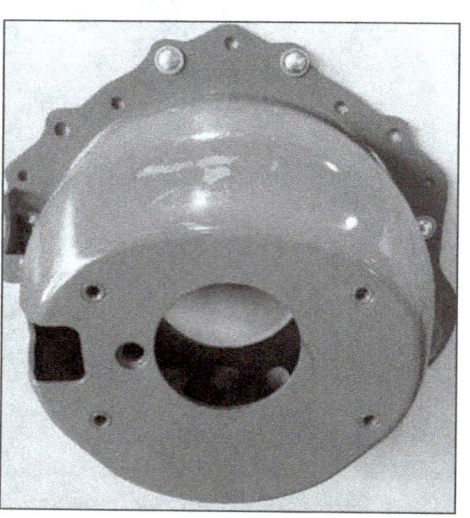
A blowproof bellhousing (or scattershield) is a valuable safety device. Many drivers have avoided serious injury because the scattershield protected them during a transmission failure. This unit is by Lakewood. I prefer QuickTime bellhousings because they typically have zero runout, and do not need offset dowel pins to make adjustments. These bellhousings are available for many unconventional, non-stock applications.

A HIGH-PERFORMANCE TRANS-FORMATION

The tunnel needs to be opened up for transmission clearance and the shifter. Only a moderate-size hole needs to be cut in the transmission hump. There is no template; you need to bolt the transmission to the engine and test fit it to determine exactly how much you should cut. You can fabricate a simple cover to allow a bit more clearance around the shifter for easier in-car maintenance.

This car is going to be channeled slightly and the floor raised in places to accommodate the exhaust and a full belly pan. Because this car is going to be used for land speed racing, we will likely cut out the tunnel entirely and replace it with new metal.

Factory bellhousings are readily available on the used market, but I highly recommend a safety bellhousing (or scattershield) to anyone who is running a manual transmission. This is considered required equipment for racing but it's smart for high-performance applications as well. A clutch and flywheel explosion can not only destroy your car, it can maim or kill, so don't consider this a "race-only" modification.

Today, some form of the 6-speed T-56 transmission is installed in the vast majority of G-Bodies, which are converted from automatic to manual transmission. Often, this is part of a GM LS-series engine swap. The *GNXcess* represents a typical type of swap.

To start, I installed a 6.0-liter LQ4 with forged internals and stroked to 408 ci. Because this was originally an automatic car, I installed the T-56 and an F-Body–based hydraulic linkage to keep things simple. I used a stock 2001 F-Body T-56 to mock up the transmission in the car, and the original flywheel, clutch assembly, and bellhousing.

All of this will be upgraded later for the installation of twin turbochargers, and the drivetrain must be as solid as possible to handle all the power (in the neighborhood of 1,400 hp).

Transmission Fit

The first thing I needed to do was establish the fitment and position of the transmission in the chassis and make any changes necessary.

Here are the general steps I followed:

1. Install the engine and transmission on BRP/Musclerods engine mounts to mock it up. The stock tunnel needs to be modified for the transmission to fit, but this isn't difficult.

2. Once you're satisfied with the fit, install the transmission crossmember (as described in Chapter 11), and jack it up into the tunnel as far as it goes. You'll be able to see the areas causing interference.

3. If you are "splitting" the tunnel, as described in BRP's directions, scribe a line down the center, and cut it with a cutoff wheel. You can then pry it up, creating clearance for the transmission, and tack it back in place.

4. Add a piece of sheet metal to fill the gap (14- to 16-gauge, cold-rolled steel). In this case, I just cut out the part of the tunnel that interfered with the transmission, and will make a new top piece for the tunnel later.

You can use the pictures of my tunnel as a guide, but check carefully to be sure you have a good fit, and don't cut too much. I'm allowing plenty of clearance for the shifter so that it can be easily removed, if needed, without dropping the transmission.

Clutch Pedal

Initially I planned to adapt F-Body pedals to the original pedal box, but I decided to use a set of pedals from an early G-Body. To make the conversion, the brake pedal needs to be replaced or modified because it is much larger than the brake pedal in a manual car. Follow these steps:

1. If you haven't completely removed the interior, do it now. Working under the dash in cramped quarters is no fun, and when replacing pedals, it's just best to get it out of your way. (This car is getting a full rebuild, so it all has to come out anyway.)

2. Unbolt the original pedal box, and either modify it with

CHAPTER 10

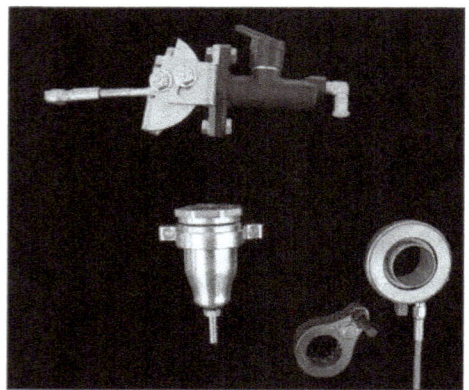

Most manufacturers offer these clutch hydraulics. The master cylinder assembly (top) is adjustable using the two bolts to suit the angle of any firewall. The reservoir is a nice billet piece, and the slave/throwout bearing assembly is a retrofit unit. LS-style T-56 transmissions use a similar unit, which bolts on like a bearing retainer. The LT1-style T-56s use a bellhousing-mounted slave with a clutch fork and throwout bearing.

This Wilwood clutch master cylinder offers another option for hydraulics. These are designed for racing, so there are many different bore sizes and mounting choices.

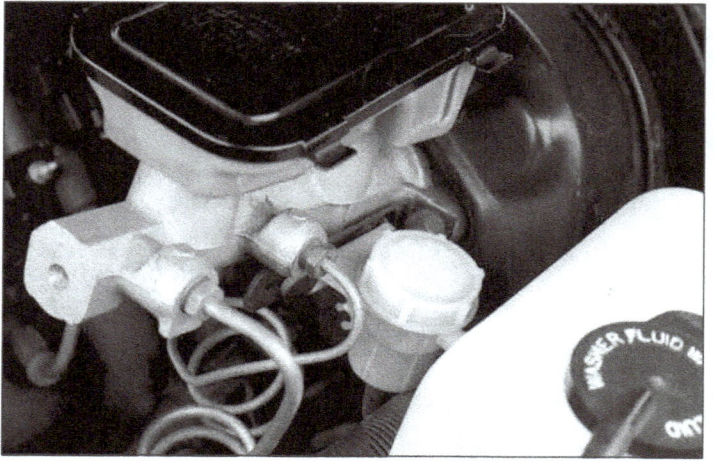

The small plastic reservoir (right) is a stock fourth-generation F-Body piece, as is nearly everything on this Monte Carlo LS.

This shows how the hydraulic slave/throwout bearing assembly is attached. It simply bolts to the front of the transmission. This is an American Powertrain unit installed on a Tremec 5-speed.

pedals from another application, or replace it completely.

3 Examine the firewall and look for a circular stamping to the left of the clutch pedal. Most cars still have the location marked, even though few manual transmissions were installed in G-Bodies from the factory. The clutch master cylinder passes through this area.

The type of master cylinder, such as Tilton, Wilwood, General Motors, McLeod, Tick Performance, etc., and the linkage determine the size of the required hole. It also determines how much modification is needed to connect to the master cylinder.

Each time you depress the clutch pedal, you need a clean, direct stroke on the master cylinder's piston, and there can't be any deflection or binding. If there is binding and deflection, the geometry isn't correct, so even though it works, it will leak and eventually fail. The firewall itself isn't strong enough to support the pressure from repeated clutch use, so the firewall must be reinforced in some way.

4 Assemble the linkage from the pedal to the master, and ensure that the amount of travel is correct for the master cylinder to achieve a full stroke.

5 With the master cylinder connected, route a line from the master cylinder to the slave cylinder, which is located inside the bellhousing. You can use a factory line and factory-style hardware or an AN-3 Teflon line.

A HIGH-PERFORMANCE TRANS-FORMATION

6 Once connected, determine the mounting location for the master cylinder reservoir. A common location is on the outer brake master cylinder mounting stud.

7 Once mounted, run the line down to the master cylinder, fill the system, and begin the bleeding process. The bleeder screw is located just inside the bellhousing.

8 Once the pedal comes up and responds normally, and there are no leaks, this part is complete.

Driveshaft

You probably need driveshaft modifications to connect your new transmission to the existing differential. This depends on what transmission you are replacing, and what you had initially. Sometimes you get lucky and don't need much. For example, a T-56 is the same length as a TH700R4 (or 4L60/65/70E), and most traditional 4-speeds are the same length as a TH350 or TH2004R. Worst case, you may need a new driveshaft or a longer tube.

Speedometer Drive

For speedometer operation, the transmission needs to be compatible with the speedometer. The T-56 and most T-5s use a VSS signal, rather than a mechanical speedometer drive. To convert from an electronic version to a mechanical version, use a Cable-X box from Abbott Enterprises, a modified tailhousing from Jaguars That Run (in the case of the T-56), or just change over to the mechanical parts from an older T-5. The choice is yours.

Shifter

A higher grade shifter, whether with external or internal rails, improves the performance of any transmission. You need a high-grade shifter for any OEM-style 4-speeds because most have a low-quality stock shifter, which has poor shifting action resulting in mis-shifts. I use Hurst shifters for older 4-speeds, such as the Competition Plus (part numbers vary according to the transmission being used; you also need the installation kit, which has the linkage).

For the T-5, T-56, and Tremec 5-speeds I use a McLeod shifter, which has a very positive feel and adjustable stops. McLeod shifters have an offset design that works well with factory consoles.

Sometimes you may need to modify (cut and re-weld) the shifter handle for better console clearance. A variety of bolt-on sticks are available from Hurst and others. This is less of an issue with a G-Body because there weren't any console-equipped manual cars, and most run without a shifter, or run a very large shifter boot and a custom-made plate to replace the old, automatic shifter and indicator. This allows plenty of room.

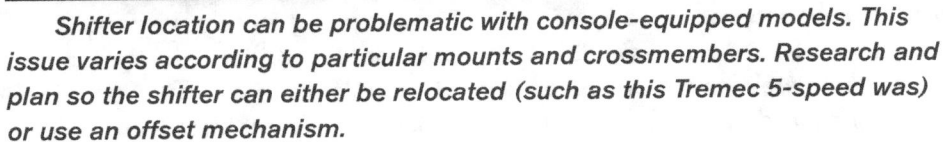

Shifter location can be problematic with console-equipped models. This issue varies according to particular mounts and crossmembers. Research and plan so the shifter can either be relocated (such as this Tremec 5-speed was) or use an offset mechanism.

If you don't want to assemble all the parts needed for a swap like this yourself, companies such as American Powertrain can complete the job for you. It offers complete Tremec 5- and 6-speed swap kits, which make for easy installation into a G-Body.

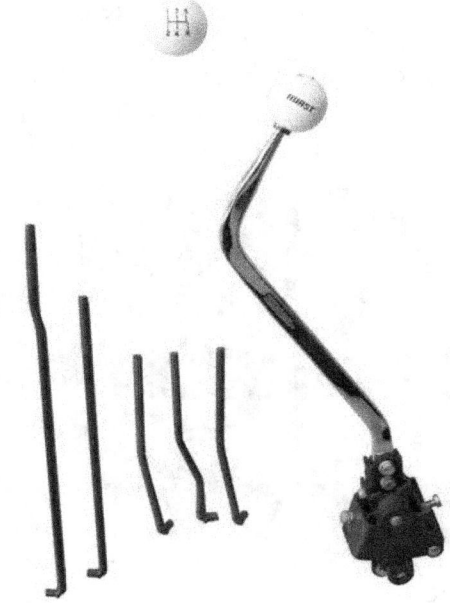

Choosing a traditional 4–speed? Go ahead and replace the factory shifter with a Hurst; the stock ones are junk, especially after some miles have loosened them up.

CHAPTER 11

A HIGHER POWER
ENGINE SWAPS

Most G-Bodies were not high-performance cars, with the notable exception of the Grand National and a few other models. Many G-Body cars were equipped with low-output engines, such as the throttle-body-injected (TBI) V-6, Chevy 305, and Olds 307. The most powerful engine by far was the turbocharged, intercooled LC2 Buick 3.8-liter SFI V-6, which was underrated at 235 hp in 1986 and 245 hp in 1987. In fact, the Buick Grand National, which was equipped with this engine, earned the distinction as "fastest American production car" for two years in a row in the mid 1980s.

However, this engine is not comparable to, or nearly as powerful as, many new-generation crate engines, particularly the GM LS engine series. But keep in mind the Buick Grand National turbo 3.8-liter was a stand-out engine. Among G-Body models, the anemic carbureted or TBI-equipped V-6s were the most common, and a variety of small V-8s were factory installed. For the era, a carbureted Chevy 305 and the lesser powered Olds 307 were competent engines. Most were happy to have RWD, V-8-powered car at all. Remember that although these are net horsepower figures, they are not comparable to the modern Chevrolet Performance crate engines.

The front crossmember allows for the installation of a variety of older and newer GM powerplants, using factory-drilled holes. The G-Body accepts the Chevrolet small-block, big-block, and of course the GM LS engines. So, you can bolt in a much larger, more powerful engine in place of the original one, without the hassles typically experienced when swapping other engines. In other words, you can easily swap your anemic Olds 260 or 307 for a 350, 400, or 455, using the existing frame stands and mounts.

In addition, you can install a Chevrolet engine into a Buick or Pontiac using all factory parts. General Motors drilled all the frames with multiple bolt patterns, which makes it easy to install any of the

The 1986–1987 Regal T-Type and Grand National came with the 3.8-liter SFI and the turbocharged and intercooled LC2, making it the fastest American street car in the mid 1980s. Few even consider swapping one out because these engines have great potential for power. With some work, the turbo V-6 can easily double the factory rating in street applications. However, there are many LS and other engines that can easily surpass the power levels of a turbo Buick.

The small-block Chevy is the most popular engine to swap into the G-Body. While many G-Bodies came with a small-block Chevy, none produced inspiring muscle car performance, and in fact, many produced less than 175 hp. As a consequence, engine swaps have been a popular first modification on these cars. (Photo Courtesy Doug Lutes)

It's fairly common to see a modern engine swapped into a G-Body, such as this 1983 Regal T-Type. The original carbureted turbo V-6 was removed, and a stock Vortec truck engine replaced it. Note that the carburetor and intake are from Edelbrock, as are the shorty headers. In emissions-controlled areas, this type of combination does not pass. However, Georgia uses a rolling 25-year cutoff for emissions in the metro Atlanta counties, so this car was legally tagged.

Ben Meissner owns this well-detailed 406 that provides plenty of motivation for the Pumkinator. It has low compression, a mild-performance cam, and home-ported stock cylinder heads. This engine is very torquey, and was a great choice for this car. (Photo Courtesy Ben Meissner)

This Vortech supercharged 350 makes a nice addition to any G-Body. It makes great power (about 500 hp at the flywheel) while still retaining good street manners. With Vortech's supplied drive, it's an easy installation.

above mentioned engines into any G-Body, regardless of which engine originally came in the car. Pontiac engines came in very few G-Bodies, and none came with Cadillac or Buick engines, so these require a bit more ingenuity when swapping.

Traditional V-8 Power

The following are some classic V-8 engine swap options you might consider. Because I cannot cover every single compatibility aspect or issue you might have to deal with, I recommend doing further research on your combination to be sure there won't be any surprises. The various online forums, such as MalibuRacing.com and Gbodyforum.com, are great resources for swap info.

Chevrolet

The small-block Chevrolet is still the most popular of the traditional V-8s. If your G-Body came with a 305, you can easily replace it with a larger small-block. All the original front accessories bolt right up, as long as the engine has accessory bolt holes in the cylinder heads.

There were some changes from year to year, though, so be sure of what you have before launching

CHAPTER 11

into a swap project. For example, if you are swapping in a Vortec long block from a late-1990s Chevrolet truck, you must use a Vortec-specific aftermarket intake manifold because the straight-down bolt pattern is completely different than that found on other engines. The engine mounting locations are the same, though, so you can reuse your existing mounts.

Also, if the old engine has a two-piece rear main seal, the old flexplate is not compatible. Use the one on the later engine. Factory engines with roller cams need a compatible distributor gear.

LT1 swaps aren't as popular as they once were, partially because there are fewer tuners in some areas, and partially because of the increasing popularity of LS engines. One alternative is to use the LT1 with a carburetor. The LT1 utilizes a reverse-flow cooling system so a standard intake doesn't just bolt on. However, Chevrolet Performance offers one that does. It is shown here with the LT4 Hot Cam, roller rockers, and cylinder heads.

The big-block Chevrolet is another swap option, but keep in mind that it adds weight to the front end of a car. In fact, the Mark IV big-block 396 is about 140 pounds heavier than a typical small-block Chevy engine. This 396 has the original A/C box, but the brake booster has been removed.

Swapping a big-block Chevy into a G-Body is relatively easy and straightforward because the big-block engine mounts match up to the small-block mount points on the subframe. Any passenger car accessories are compatible, and most factory exhaust manifolds fit.

To install a small-block Chevrolet engine in a G-Body that wasn't originally equipped with one, you need frame stands from a Chevy-equipped car and a compatible transmission with the correct bolt pattern.

If installing an LT1 or LT4, use standard Chevrolet G-Body frame and engine mounts, and keep the factory accessories from the LT1 because the layout is completely different than a standard first-generation small-block.

The G-Body's engine compartment is large enough to accommodate a variety of engines including the big-block Chevrolet. Chevrolet never installed a big-block in these cars, but big-block engines align with the existing small-block mounts. No adapters or conversion mounts are needed. In the mid 1980s, I attended Super Chevy events and saw half a dozen, brand-new Monte Carlo Super Sports with LS6 454 crate engines installed that looked like factory installations. Most even used the feedback Quadrajet and computer-controlled ignition.

Most passenger car big-block exhaust manifolds fit, but headers produce much more power and many are available for a variety of setups. Hooker makes a set (PN 2241) with 2-inch primaries and 3½-inch collectors that fit very well and even clear factory clutch linkage. V-belts are ideal for a big-block passenger car with accessories and the long water pump (1969-newer), but a bracket for the shorter, radial-style A/C compressor must be built. An easier solution, especially if your G-Body is a 1985 or newer with a small-block Chevrolet, is to retain the existing serpentine accessories. Kwik Performance can provide the brackets, making the swap very simple.

A HIGHER POWER

Cooling is more of an issue with big-blocks, so this is a good time to upgrade the radiator. The stock radiator is not sufficient, especially with air conditioning.

Hood clearance is another important factor with a Chevy big-block swap. With the stock hood, you need a very low rise intake for enough hood clearance, and even at that, you still need a dropped-base air cleaner. When installing a Chevy big-block, run a taller hood, such as a cowl-induction type.

Buick, Oldsmobile, Pontiac

All owners do not install Chevy engines into their G-Body cars,

Rick Bejarano of METCO Motorsports chose a 502 crate motor from Chevrolet Performance for his Monte Carlo SS. It is backed up by a Tremec 5-speed. While definitely heavier than a small-block, the aluminum heads lighten things up a bit. The instant torque makes this a fun car to drive.

Jeremy Hale's Cutlass came with a 3.8-liter V-6, but he swapped in this Oldsmobile 350. It's internally stock, but with the Edelbrock intake and headers it runs very well. It was never dyno tested, but horsepower and torque are estimated in the low 300s, which isn't a lot, but considerably more than any G-Body came with from the factory. It's relatively easy to install an Olds 400 or 455, and they have more potential than the 350.

This Olds 350 came out of the Lonestar, a 1965 Cutlass; a 525-hp crate LS3 is going to go in its place. The horsepower advantage of the LS3 is obvious, but even at that power level the drivability and mileage is comparable to a new Corvette. In spite of its rough looks, this 350 engine ran well, and is now swapped into a 1986 Cutlass.

This sinister-looking Buick Regal doesn't have the typical turbo V-6, or even a Buick 455. Instead, a built Cadillac 500 has been installed. While a little tougher than some swaps, the engines are inexpensive and still plentiful, and they have good aftermarket support. They are also lighter than most Buick, Olds, and Pontiac engines and make more than 500 ft-lbs of torque even in stock form.

A 500-ci Cadillac from a 1968 to late-1970s RWD fits a G-Body using factory-style mounts. In this particular build, an engine plate has been installed.

Most gearheads overlook Cadillac engines, but the 472s and 500s produce 500 ft-lbs of torque in stock form. With an aluminum intake manifold, they weigh about 575 pounds, which is the same as a stock, all-iron small-block Chevrolet. This one isn't stock, and sports Wenzler aluminum heads along with the typical bolt-ons.

especially Buick, Oldsmobile, or Pontiac cars. BOP makes actually have things a little bit easier when it comes to using a larger-than-stock engine because all Oldsmobile and Pontiac engines are based on the same architecture, so dramatically increasing cubic inches is much easier.

Buick small-block 350s used different mounts than the big-block 400 and 455. While Cadillac never made a G-Body, owners have installed the Caddy 472 and 500 V-8s in G-Bodies. To complete the swap, you need mounts from a RWD Cadillac (any 368/472/500) and the rear sump pan from an Eldorado. Headers can be a problem, but some Cadillac specialists offer Cadillac-specific header flanges that can be used to modify an off-the-shelf Chevy big-block header or as a basis for a set of custom headers.

The Olds V-8 engine of your choice bolts up to the 260 or 307 engine mounts and accessories. When using factory exhaust manifolds, the A/C box may need some slight trimming at the bottom, but most owners typically install headers, so this isn't required anyway. When swapping a Pontiac V-8, use the mounts from an early 301-equipped G-Body (rare), or get the mounts from any Pontiac RWD V-8. Buick V-8s were not installed in the G-Body because these engines were already discontinued when the car was released. But just about any Buick engine and frame mounts that fit the earlier V-8 Buick engine work. However, in most cases, you need to drill the frame for the mounts.

LS Series

There are many GM LS engines available, and there's one to fit almost any application and budget for a G-Body. The main thing to keep in mind when contemplating one of these swaps, however, is to be sure that the parts you are using are compatible with one another. For example, be sure that oil pan clearance, accessory drive clearance, reluctor wheel changes, accelerator cable versus drive-by-wire, and gauge hookup are compatible with your car and setup.

The list of possible swaps can go well beyond what you find here and may seem nearly endless to the first-time swapper. To make things

An LS engine swap into a G-Body requires a custom oil pan. The stock pan must be replaced with one like this. BRP/Musclerods calls it the LH8, but essentially it is a Hummer pan with gasket, windage tray, dipstick and tube, and all the needed bolts. BRP's version has some extra machining done for the AFM (active fuel management) engine. This system turns off half of the engine's cylinders for improved emissions and mileage at cruising speeds, but if you don't require that feature (most don't), get the Musclecar Pan Kit from Chevrolet Performance.

A HIGHER POWER

The LS1 engine in this Malibu has been set up with a return-style fuel system (note the regulator on the firewall), rather than the returnless system used on most swaps. If you're planning to make more than 550 hp, and you're using larger AN plumbing (-8 or larger), then this type of system makes sense.

simpler, you can often purchase an engine and transmission as a package, along with the original ECM, wiring, accessories, and pedal (if drive-by-wire).

The GM Gen III/IV engines are the most popular to swap into the G-Body chassis. These engines are sometimes known among enthusiasts as "LSx." Note the small x, to avoid confusion with GM's race-oriented LSX engines, which are based on the same general architecture. Some simply call them all "LS" engines, even though not every one had an RPO code that started with LS. An entire book of the various available versions and the common modifications could be written, but here I focus on the basics of these engines as it applies to swapping them into the G-Body.

LS1/LS6/LS2

The all-aluminum LS1 was first released in 1997 in the C5 Corvette and was the first engine in the GM LS engine family. The next year, it was also offered in the Camaro Z28, the Pontiac Formula Firebird, and the Pontiac Trans Am. In addition, the LS1 was installed in the 2004 GTO. The Camaro/Firebird versions are the most common.

A higher performance version, the LS6, was installed in the Z06 Corvette (C5) and the Cadillac CTS-V (first generation). All used pressed-in cylinder liners, such as all GM LS-series aluminum blocks, and therefore the liners can be only slightly overbored, rather than bored similar to a traditional iron V-8.

If necessary, sleeves can be replaced. Aftermarket sleeves are available to stretch the displacement beyond the factory offerings, but they are pricey. In most cases, if the block needs to be re-sleeved, it is cheaper to find another engine.

LS1 cylinder heads have an unusual cathedral-port design, which flows extremely well, and is largely

If you cannot afford a crate engine, consider a low-mileage take-out, like this one. This particular LS2 engine and 4L70E transmission powered a 2009 Trailblazer SS with fewer than 20,000 miles. Purchased as a package deal, it has the Corvette intake, a new harness (designed for the application), and a PCM reflashed by Travis Digby.

Mike Meadows' LS1 features Trick Flow heads, a Vengeance Racing custom-grind camshaft, an Edelbrock Pro-Flo intake manifold, and nitrous. It makes about 475 hp at the wheels without the nitrous. A combination such as this makes the car an incredibly fast street machine. A pro-touring G-Body car could easily compete with this combination at national events. The accessory brackets are by Kwik Performance, the preferred brand for a typical LS swap.

responsible for the power potential of these engines.

The LS2 followed, and was essentially a 6.0-liter LS1.

LS3/LS7 Aluminum Blocks

When the LS3 debuted, it looked very similar to the LS1 and LS2 but displaced 6.2 liters. A more traditional rectangular port replaced the familiar cathedral ports, which are better suited for higher RPM power and larger displacement. They are found stock in the Camaro SS and the new Corvettes. The LS3 rapidly became the most popular engine for those wanting to use a new engine for their swap because Chevrolet Performance priced crate versions very competitively, and even developed their own harness and controller package.

We installed a 480-hp version of the LS3 in the *Grocery Getter* (the latest offering is rated at 525 hp). For those needing an emissions-friendly package, an E-Rod version is an option, and it is a fifty-state legal swap. Keep in mind that there are also truck versions of the aluminum 6.2, which feature similar L76 rectangular port heads.

The LS7 first appeared in the C6 Z06 Corvette, and in many ways, it is the ultimate naturally aspirated LS engine, but it is expensive! Because they sell for almost twice as much as the LS3-based crate engines, many don't have the budget to install one in a high-performance street build. It's most often seen in track cars and high-budget buildups. It has rectangular-port heads, but with even more flow than the LS3 versions, and 7.0 liters of displacement: 427 ci. It also has titanium connecting rods, a dry sump oiling system, and a few other features more often seen on race engines.

The LSA and LS9 engines are very similar. Both are supercharged and based more on the LS3 than the LS7. If you want a factory-supercharged engine with a warranty, this is a great way to go, but they are very expensive.

LQ4/LQ9 Iron-Block Truck Engines

Trucks came with iron blocks and aluminum heads in 4.8-, 5.3-, and 6.0-liter displacements. These engines weigh a bit more than the all-aluminum engines (in some cases 75 to 150 pounds), and that weight on the front end does not enhance handling. But these engines also add durability, and can be easily overbored if necessary, rather than using the expensive cylinder sleeves of the aluminum blocks.

The most popular truck engines are the LQ4 and the higher compression LQ9 6.0-liter engines; but buyer beware, many 4.8- and 5.3-liter engines are being sold as 6.0-liter engines. That said, a 5.3 or even the 4.8 is worthy of high-performance applications and respond very well to modifications.

Chevrolet Performance offers a variety of crate engines, and the 525-hp LS3 provides more dollar per horsepower than many others. It's a veritable bargain at around $7,500. The LS3 isn't equipped with accessories. Some equipment, such as the oil pan and exhaust manifolds, have to be discarded because these production Corvette parts aren't compatible in the G-Body chassis.

This highly detailed LS7 has a custom sheet-metal intake and carbon-fiber accents. The dry sump oiling system has been retained; the tank is located at top right. The car is also equipped with Hydroboost. This is an early Corvette, not a G-Body, but any of these features could just as easily be used on a G-Body.

With any of these engines, a larger camshaft, new valvesprings, and a set of headers can significantly increase torque and horsepower. The later 5.3s and 6.0s are aluminum, and for this reason positive identification is a must because the 6.0-liters warrant a premium price. Most of the truck cylinder head castings respond very well to porting and other modifications; they share the same cathedral-style ports as the LS1 and LS6 heads, and often the only major differences are the size of the combustion chambers (typically larger, for slightly lower compression) and the size of the valves.

In many cases, the casting numbers are shared with the LS1 engines and any LS-based factory cylinder head, except for the LS3/L76 and LS7 rectangular port heads that can be used on the 4.8 and 5.3. The 6.0 can use any factory head.

Keep in mind, though, that as with most other engines, the piston volume and combustion chamber volumes vary from one version to another, so take this into account to be sure you wind up with a favorable compression ratio. Typically, the lower performance truck engines come with larger cc heads that are excellent for dropping compression on an LS1 for blower or turbo usage.

Engine Mounts

A multitude of engine mounts are on the market. The vast majority are "plate" style, which is essentially a steel plate that bolts to the block. In most cases, plate-style mounts are compatible with the original V-8–style frame mounts. Some are adjustable, and they all place the engine in a slightly different position in the engine bay.

Some claim to put the engine exactly in the stock location, but in many cases, it's not desirable to place an LS in the stock engine location. Because the engine is not the same shape and size as the original, this can cause problems with the oil pan, steering linkage clearance, driveline angle, and others.

If you insist on keeping the mechanical clutch linkage, it may be worth it to mount it in the stock location, but I can't think of any other valid reason to do so. Regardless of position, many times the oil pan contacts the front crossmember, so an expensive aftermarket pan or pan modification is required for clearance. In addition, steering linkages often rub the pan at full lock, even with an aftermarket pan. Driveline angles are often incorrect, resulting in vibrations, and the driveshaft can severely deflect; the result is a jump-rope effect. In these extreme cases the driveshaft, U-joints, and other parts can break.

However, all of these issues can be corrected with enough ingenuity, but they take time, and you may not have that time if you need the car for daily transportation.

Once those issues are resolved, you need to find headers that fit with the mounts. Some manufacturers, such as Hooker, Stainless Works, and Kooks, offer "swap" headers that are designed to work with (or just happen to fit) a particular mount. Unfortunately, choices of header type and size are still very limited. For instance, a mild 4.8 has very different requirements than an LS7, so this is an area where compromise is usually necessary.

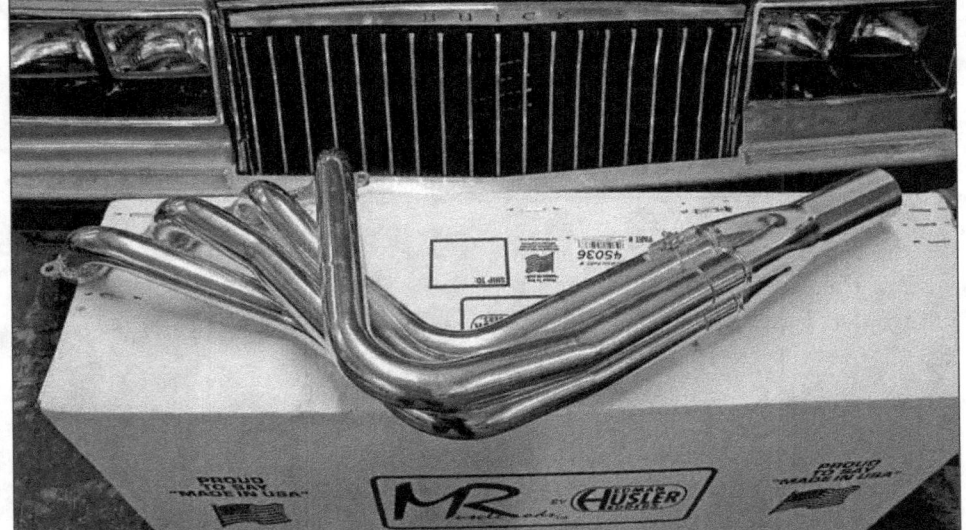

BRP/Musclerods offers headers in either mid-length or long-tube styles and a variety of sizes that include stepped versions. All come with a chrome-like ceramic coating, gaskets, bolts, and oxygen sensor bungs to be welded into the exhaust. The collectors can be unbolted for the heads to be installed in 30 minutes, even with the engine already bolted down. The slip-fit at the rear of the collector eliminates the possibility of blown gaskets, so you just add some high-temperature RTV, and slide the head pipe into place. A simple stainless-steel band clamp keeps it secure and leak-free.

CHAPTER 11

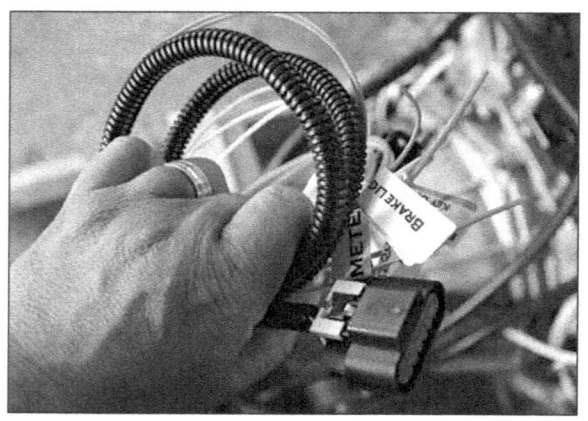

Wiring is an intimidating step for most people. But a pre-made harness, such as this one from Modern Vintage Systems, is very easy to install. All the connections are labeled, and most only fit one connector anyway, so there is little possibility of error. The loose wires pictured are the only ones that go to the car itself.

These Holley Performance Products harnesses and controllers are for the G-Body; MSD, F.A.S.T., and others offer a similar system. The many self-tuning systems take a lot of the guesswork and expense out of engine tuning.

Wiring and Engine Management

Wiring and engine management doesn't have to be an intimidating part of the swap, as long as you are realistic about your abilities and your budget.

Option 1

Most swappers use the factory harness, ECM, and drive-by-wire pedals when installing takeout engines. It's the least expensive method, but it can also be a major headache. To make the harness compatible, pare it down to exclude any unneeded functions. Some wires need to be lengthened while others need to be shortened. The harness usually needs to be repaired in some way. Often, an accident or careless person pulling the engine damaged or inadvertently modified the wires and/or connectors.

If the harness has been modified, it needs to be re-sleeved or loomed for protection. Instructions and pin-out diagrams available online show you how to do this, and it's fairly simple. If you don't have a lot of patience or a talent for wiring, you may want to consider having a pro do this. Most charge about $350 to $450 for the harness modifications. Another $250 to $300 is charged for deleting the unneeded emissions and anti-theft functions from the ECM and adding a basic performance tune.

I prefer to use a new harness, which is custom tailored to the particular engine swap. I typically buy a harness from John Leonard of Modern Vintage Systems. He builds them to my specifications so the harness is precisely constructed for the engine and chassis combination. Thus, it's a clean and precise fit, which ensures a far neater job. The ECM is placed out of sight, and the harness is routed in a way so it is safe from damage.

The cost is higher than a modified original, at $700 or so, but the peace of mind and enhanced appearance of the new harness is well worth it to most. Modern Vintage Systems also does the programming, so there is less chance of error in matching the programming to the hardware. This is very critical, as all drive-by-wire pedals are not the same.

Option 2

Installing a Chevrolet Performance harness and controller package is another alternative when installing a GM LS crate engine. Controller packages are available for most of these newer engines, and they come with a custom tune that works great right out of the box.

GM controllers and harnesses for the LS376/480-hp crate engines with a crate 4L70E were installed in the *Grocery Getter*. They performed beautifully and allowed excellent control of the crate engine and transmission. The kit includes oxygen sensors, sensor bungs, a drive-by-wire pedal (ours had one from a Cadillac CTS), an MAF insert, and several other items that aren't typically included with a harness.

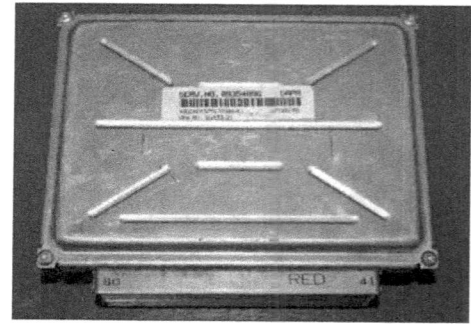

This LS1 PCM has been reprogrammed, with all non-essential emissions and anti-theft functions eliminated. Older units like this one control both the engine and transmission; later models use a separate controller for the transmission, but are much smaller.

Option 3

A true aftermarket controller system from F.A.S.T., Big Stuff 3, Holley, and others is another option. These systems are infinitely tunable so the system can be set up for special fuel requirements and applications. On the flip side, you may not be able to properly set up a system if you're not a proficient tuner.

These speed density based systems can help make immense power, but they don't adjust for varying altitude or atmospheric conditions as a factory MAF system does. In other words, they are great for 1,500-hp drag cars, but not as great for a car you are building to take on the Hot Rod Power Tour. As conditions change, the tune is off.

General Motors uses a MAP and MAF sensor combination, to ensure greater accuracy and the ability to adapt to changing conditions, such as temperature, humidity, and altitude.

Exhaust

Installing aftermarket exhaust systems in a G-Body for an LS engine swap is actually easier than installing the typical conventional Chevy small- or big-block engine exhaust. For example, BRP/Hedman headers use a slip-fit design, so installation is much easier. The 1¾-inch-long tube headers are suitable for a 2½-inch system; while the 1⅞-long tubes are a good match for a 3-inch system.

Instead of having to mess with a three-bolt collector flange, or a ball-and-socket design, you can simply slide the head pipes into the collectors. I use a little copper RTV on them and clamp them in place with stainless band clamps to avoid crimping the pipe.

You can take an appropriately sized dual-exhaust system from a late-model A-Body and adapt it to your G-Body car. (Magnaflow and Flowmaster systems are good choices.) You probably have to make some modifications for a particular kit to fit your car's body style. For the *Grocery Getter*, we used a Chevelle wagon system that had 3-inch exhaust tubes. The exhaust was routed to exit under the rear quarter panel, which took a little cutting and splicing with a Magnaflow kit.

Alternatively, you could have a muffler shop build a system for you, but in most cases, they won't have a mandrel bender. Crimp bends are adequate for a daily driver, as long as they are done properly, but I prefer to stick with mandrel bends.

Intake

As great as the power potential is with a naturally aspirated LS-series vehicle, a surprising number of swaps incorporate some form of supercharger (belt driven) or turbocharger (exhaust driven) for even more power. Power adders such as these are a great way to get significantly more power (double or more) out of your engine, while retaining good drivability and mileage. Getting 500 hp out of the smaller, less radical (low-boost) setups isn't unusual, and those willing to push the envelope a bit can easily exceed

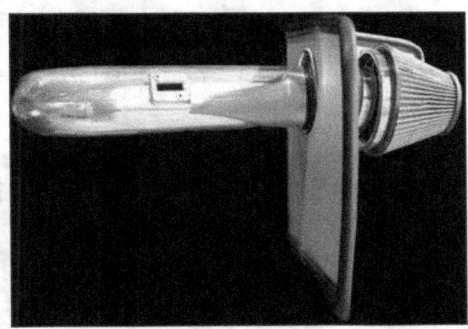

Spectre Performance makes cold-air induction systems for late-model applications, and recently released a hot rod system designed for retrofits. Spectre also sells every piece separately, so you can build whatever you need. If you're using a later engine with a MAF insert, the late-Camaro SS system works very well on a G-Body.

The engine for Holley's G-Force One *is impressive. It is based on an LS7 block with an LS9 supercharger. This setup would easily cost $20,000 or more to duplicate, though, so it isn't for budget builds.*

This El Camino has an 88-mm turbocharger feeding an internally stock LQ4.

Similar in design to an LS9 supercharger, this Magnacharger provides enough boost to make 750 hp on a well-prepped engine like this one.

The owner of this Malibu added a mostly stock truck engine, and fabbed up a twin-turbo setup. Judging by the compressor size, low-end torque and drivability are likely excellent.

Note the high-heat wrapping on the turbo plumbing. It works well on stainless tubing, but can lead to metal fatigue and cracking in mild-steel applications.

1,000 rear wheel horsepower. Some manufacturers offer kits for swap applications, but most owners create home-built systems.

Factory air-intake plumbing is very restrictive and bulky. In most cases, it isn't easily adapted to a swap. One option is to build a complete custom system, but that requires steel or aluminum mandrel bends, and all the miscellaneous connectors and clamps to put it together. You can certainly do that, but I have found an easier way, and it doesn't involve hanging a little billet filter off the end of the throttle body, street rod style.

Cold-air induction produces a stronger air/fuel charge; electric fans can have a detrimental effect blowing on a MAF-equipped vehicle's air filter. You can now get off-the-shelf parts that didn't exist when the first swaps were being made.

These days it is as easy as doing a product search on the Summit Racing website under Spectre Performance. Spectre used to be known for chrome doo-dads and fake braided hoses, but in recent years the company has built really good cold-air intake products for carbureted and EFI applications.

I first used a Spectre intake on the *Grocery Getter*. Originally, a custom-built air intake was planned for this car, but a very tight deadline forced us to go another route. Fortunately, I determined that Spectre's 2010–2011 Camaro SS intake system was roughly the right shape and length for the underhood space available, and it fit like a glove. It even had the proper mounting points built into the tube for the MAF insert that came with the Chevrolet Performance harness/controller package. Note that this pipe does not work if you are using a different type of MAF, such as one that is molded into a plastic housing.

Summit now carries the whole Spectre line, and it is possible to buy all the parts to custom build a system (for about the same price as a kit). Spectre also has a "hot rod" kit that comes with various extra parts to make it work on most swaps.

This LS3 is set up for a carb, using the original coil packs and an MSD 6LS controller. KRC Power Steering makes the breathers and accessory drive, including the power steering pump.

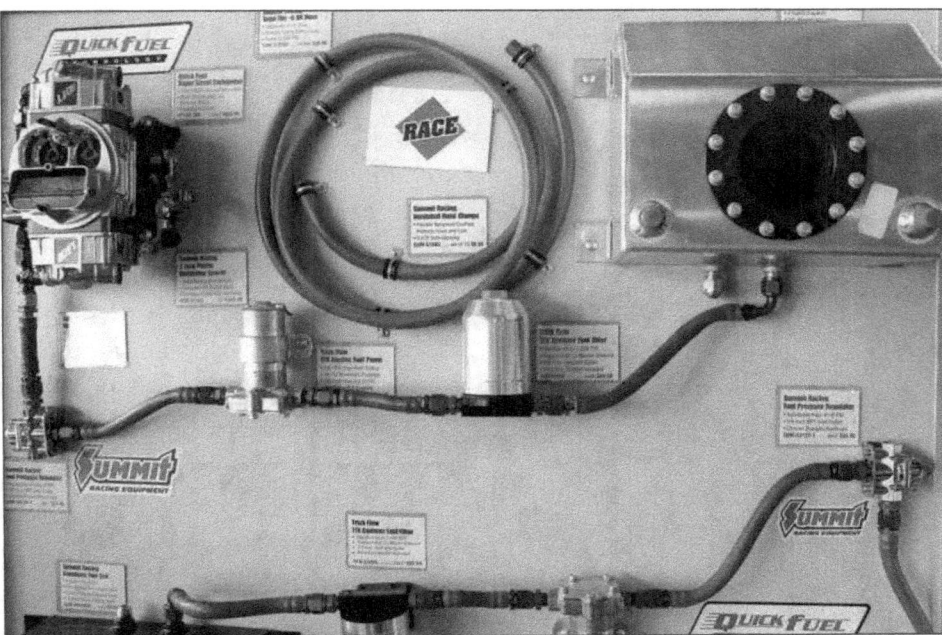

Some assume converting from EFI to a carbureted system on an LS engine is less expensive than using EFI. However, if you bought an engine with the EFI intact, it usually isn't. You still need to plumb an adequate fuel system for a carbureted LS engine, such as this example from QuickFuel Technology.

Carburetor or EFI

While some may wonder, Why not just use a carburetor and avoid all the hassle and expense of installing an EFI system? There are definite benefits to a carburetor and to fuel injection; you just need to determine your budget, performance goals, and your acumen for tuning. In my experience, most people who criticize electronic fuel injection for the difficulty of tuning can't tune a carburetor either. But, it is undeniable that using a properly tuned carburetor can provide excellent drivability, and even a bit more power, plus it is considerably less expensive than a fuel injection system if you are starting from scratch.

Holley, Chevrolet Performance, and Edelbrock all make intakes for the LS-series engines that accept a carburetor. Holley even makes dual-quad medium-rise and tunnel-ram intakes for dual-quad carburetors to supply enough fuel for large-cubic-inch engines. These aluminum intakes are more prone to heat soak than the original composite intakes, but the cooling effect is better with aluminum, so the fuel more effectively atomizes.

Some provision needs to be made for the ignition, since there is no ECM. I prefer to use an ignition controller, such as the MSD 6LS (for 24-tooth reluctors) or MSD 6LS-2 (for 58-tooth reluctors), to run the factory coil packs. They have a built-in two-stage rev limiter, nitrous retard, and adjustable advance curves. Some Edelbrock intakes come with a similar unit, and Mallory and others have their own versions.

If you have an engine that is missing the coil packs, but want to just use a distributor, Chevrolet Performance carries a replacement front cover that accepts a conventional distributor. Interestingly, it uses a Ford-style distributor.

Fans of Stacey David's Gearz TV show recognize this engine with Holley dual-quad intake because it was installed in the 1973 Malibu wagon featured on the show. If you want a more nostalgic look, this is a good way to go.

Chevrolet Performance originally designed this cover for circle track racing classes that require a distributor; otherwise, it is a solution to a non-existent problem. Unless you're racing and must have this cover, I don't recommend installing a distributor because by the time you buy the cover, an aftermarket distributor, wires, and a box to run the distributor, you could have easily bought a set of new coils and the aforementioned MSD 6LS. In other words, it's not worth it. Instead of having plug wires that are a few inches long, and that are nearly impossible to burn on the headers, you now have eight new potential places for a misfire.

If you want EFI, but don't want the late-model, plastic-intake "look," there are plenty of options. Holley wisely made originally designed carbureted intakes available with provisions to accept fuel injectors. This way, you can have a classic look and all the advantages of EFI.

Often, some who are adamant about running carburetors also want to use non-electronic automatic transmissions because they have invested substantial amounts of money in the transmissions they already have. This is understandable because the electronics to run them can be pricey. Any of the traditional non-overdrive transmissions, such as the Powerglide, TH350, TH400, or the earlier TV-cable overdrives TH2004R and TH700R4, bolt right up to the bellhousing flange. Note that one bolt hole in the transmission bellhousing isn't used (the hole is missing on LS-series engines).

With such a conversion, the spacing of the flexplate has changed, and it is impossible to bolt up the

This LSX 454 has Holley's Modular Hi-Ram dual-quad intake, equipped with black, hard-anodized carbs for a sinister look. This is a very powerful and reliable engine for a drag-oriented G-Body but lack of hood clearance makes it impractical for daily use.

An LSX 376 from Chevrolet Performance makes a nice powerplant for just about anything.

This version of the Holley dual-quad intake is meant for EFI, and comes with the fuel rails shown. The throttle bodies, as well as the entire engine management system, are from Holley. This is the School of Automotive Machinists' 2012 entry into Holley's annual LS Fest Engine Swap Challenge.

Holley puts these tags on each harness connector or wire, which makes it very quick and easy to install.

Holley provides an EFI system with a classic look: big-block–style coil-covers, 4-barrel intake and throttle body, and, of course, Holley engine management.

Hardcore racers, and those wanting a more intimidating street presence, can opt for an EFI version of Holley's Hi-Ram intake.

torque convertor and maintain the correct connection with the transmission's front pump. To solve this issue, Chevrolet Performance offers a kit, which includes a spacer for the nose of the torque convertor, spacers to go between the convertor and the flexplate, and the needed bolts.

You can perform the same conversion with an EFI-equipped engine, but you need to use one equipped with the old-style cable-driven throttle body. Adapters to do this are available from Lokar and Street & Performance.

Fuel System

Swapping a fuel system into a G-Body intimidates some enthusiast installers, but in reality nothing could be simpler, especially with the returnless fuel system found on most LS engines. The term "returnless" fuel system is misleading because there *is* a return line. In this case, it can be very short because it only has to go from the tank to the fuel filter/regulator assembly that is typically mounted in the rear of the vehicle, very close to the tank.

This system has many advantages, including lower cost (because not as much line is needed), less heating of the fuel (because the excess no longer goes through the engine compartment), and less wear on the pump. High-horsepower engines and, particularly, engines with nitrous oxide, superchargers, or turbochargers work best with conventional regulators. The vast majority of street-driven vehicles perform

You can install a fuel filter/regulator using AN -6 adapter fittings. You need a Russell #640940 -6 to 3/8 SAE quick-disconnect male fitting for the front of the regulator and a pair of -6 adapter fittings for the rear. You also need a Russell #644113 for the 5/16-inch return side, and a Russell #644123 for the 3/8-inch pressure side. You flare the hard lines with a 37-degree AN flare, using tube nuts and sleeves to connect the line to the fitting. (Photo Courtesy Matt McKahan)

The Corvette fuel filter/regulator assembly is used in many swaps because it greatly simplifies the fuel system and saves money. The ends are standard GM 3/8- and 5/16-inch fuel line flares. This assembly provides many options for plumbing the system.

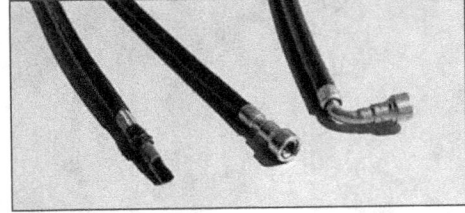
Will Handzel and the guys at TechAFX offer custom-built fuel hoses that are ideal for a variety of G-Body builds. They build hoses to order with OEM-style quick-disconnect fittings.

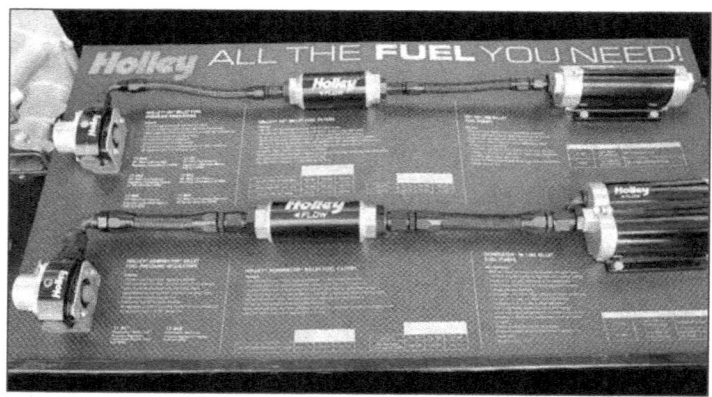

Planning something a bit more serious? Holley has you covered with a wide variety of pumps, filters, and regulators. These parts are universal to any G-Body.

well with the returnless system. For plumbing, some use the original-style plastic factory lines (Dorman makes the parts, available from any auto parts store) but I prefer to use AN-style fittings and lines.

The fuel pump is another factor to consider when setting up a fuel system. The high pressures of an EFI system require an electric pump; keep in mind the vast majority of G-Body cars were carbureted and therefore fitted with simple mechanical fuel pumps. Some turbocharged Buick Regals and the 4.3 TBI late-model

Matt McKahan sources many fuel system fittings from McMaster-Carr, an industrial supplier that is known for having nearly everything. (Photo Courtesy Matt McKahan)

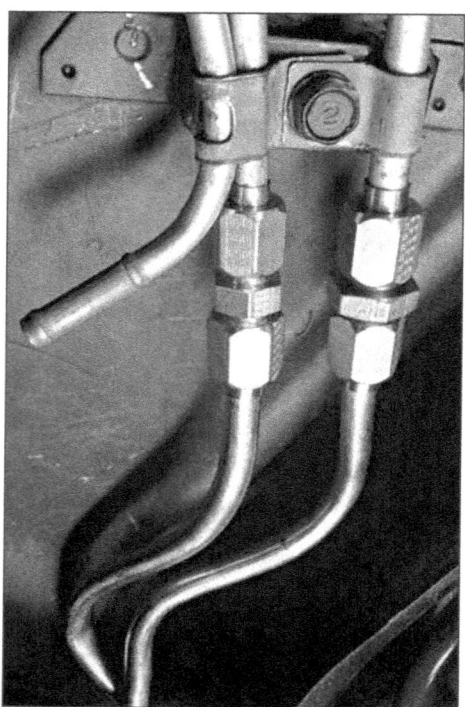

Depending on your build, you may want to use separate sections of tubing, rather than a continuous tube, because it is easier to plumb the fuel system in sections. These stainless lines have 37-degree flares with tube nuts and sleeves and are connected by -6 unions. (Photo Courtesy Matt McKahan)

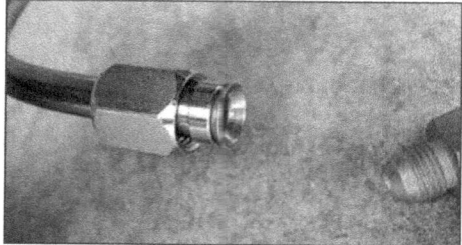

Install the tube nut and sleeve before making the flare. (Photo Courtesy Matt McKahan)

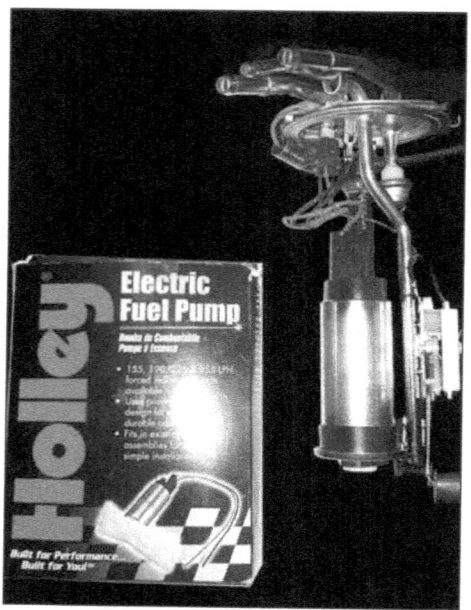

This Holley 255LPH pump is a direct fit into the Buick Grand National–style pump hanger assembly. The assembly is available from Rock Auto (PN FG156A), which is sold without a pump. (Photo Courtesy Matt McKahan)

This fuel tank is fully assembled and ready to bolt into the car. This part is plumbed with OEM-style, high-pressure fuel line and clamps. Or you can use -6 hose and AN fittings. Either works fine, and is completely a matter of personal preference. (Photo Courtesy Matt McKahan)

Monte Carlos were fuel injected and equipped with an electric fuel pump. Many swappers use in-line aftermarket pumps.

For many years, high-horsepower EFI cars were equipped with in-line pumps because in-tank high-capacity pumps weren't available. While an in-line pump is easier to change than one located inside the fuel tank, a properly installed in-tank pump that isn't abused (by running out of gas repeatedly, or keeping the fuel level low all the time; both burn up the pump) lasts far longer than a typical in-line pump. In addition, an in-line pump doesn't have the benefit of the cooling effect of gasoline submersion. An in-tank pump is also safe from road hazards that could damage it.

Installing an in-tank pump in a G-Body is very simple. You can install an unmodified Buick Grand National fuel pump/sending unit assembly in an existing tank for any coupe or sedan body style. These are available new from local parts stores. I buy them without the pump and add an aftermarket 255-lph pump. This is more than adequate to feed most any naturally aspirated engine. The bolt-in assembly fits any G-Body tank, except wagons and the El Camino/Caballero. On those models, you can still use the Grand National assembly, but the feed line needs to be lengthened because the tanks are deeper. This was done on the *Grocery Getter*, and it worked well.

Installation of the fuel system is not difficult, but specific parts are required. Here are the steps I follow:

1 Make sure the fuel level in the tank is low before starting this swap, or use a siphon or pump to remove as much gasoline as you can. This makes tank removal far easier, and it's much safer because you reduce the chance of spilling gas on yourself.

2 Once the gas is disposed of, place a jack under the tank and remove the carriage bolt on each side, which holds the tank straps in place.

3 Drop the tank a bit and disconnect the fuel and vapor lines from the sending unit.

4 Disconnect the tank ground wire.

5 Drop the tank farther and pull it out from under the car for inspection.

6 Remove all fuel lines from the car at this time and discard them. Most of the lines are held in place with steel clamps that are bolted to the frame. If the body is on, the line may have to be cut in places to get it out.

7 You can reuse the tank if it has no major dents, rust, or leaks. If it needs replacement, any parts store can provide a replacement. To remove the sending unit, carefully tap the retaining ring in a counter-

The fuel system should be run to clear all components, such as exhaust, suspension, and other parts. It must be securely clamped to the frame at regular intervals. (Photo Courtesy Matt McKahan)

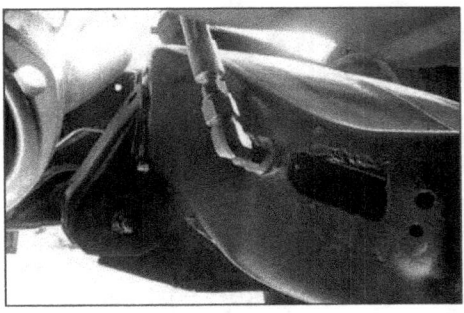

If you don't box the frame, route the hard line through the frame rail and bring it out in the engine compartment. A bulkhead fitting and a pair of 90-degree ends allow you to accomplish this. (Photo Courtesy Matt McKahan)

Tech Tip: Fittings for Fuel Hoses

Bulldawg Musclecars installs AN-flared hardline and braided hose on all fuel systems in customer cars. I prefer Earl's Pro-Lite hose because it is lighter weight than steel-braided hoses, looks good, and is easier to work with than conventional steel-braided hose. AN -6 (3/8 ID) is sufficient for most applications. If you use an OEM-style sending unit/fuel pump assembly it must be modified for AN-style fittings. ■

clockwise direction. A hammer and brass drift or punch work well for this because you don't want any sparks from a steel tool. Once the tank is loose, remove it and pull out the sending unit.

8 To install the main feed line, cut off the existing flare, and install a -6 tube nut and tube sleeve.

9 Use a 37-degree flaring tool to flare the line, and then screw in a -6 male union. This allows a standard female -6 hose fitting to be used on the flex line. The return line is 5/16 inch.

The tube nut and sleeve method doesn't work here because it is a 5/16-inch line and a 3/8-inch fitting (all sizes come in increments of 1/16 inch). Fortunately, most fitting companies make a compression-style adapter fitting that goes from 5/16 to -6.

10 As with the feed line, cut off the existing flare first and then install the two halves of the adapter fitting and place the supplied compression ferule in between.

11 When the fittings are tightened, they form a reliable seal, but the fittings must be tightened properly and the tubing must be the correct size.

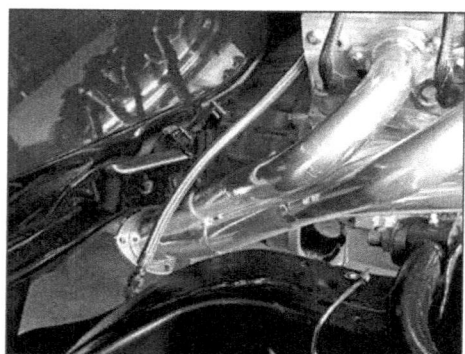

Always run a flexible line from the frame to the engine. Be very careful when routing the line; clamp it securely in place. If the line passes anywhere near the exhaust, as it does here, use a heat sleeve or Thermotech-style heat tape; otherwise, you risk a catastrophic fire. Note how well the BRP/Musclerods mid-length headers fit on this Regal's LQ4. (Photo Courtesy Matt McKahan)

The original tubing on the fuel rail has been flared and AN fittings were used. My preferred method is to use a Russell fitting or a TechAFX line with a factory-style quick-disconnect fitting. (Photo Courtesy Matt McKahan)

12 Install the new pump into the holder after both flares are completed and -6 unions installed.

13 Be sure the hose connections are tight, and then hook up the electrical connections to the pump. Depending on what parts you are using, you may need to change the terminals.

14 Install the new filter sock (included with the pump) on the end of the fuel pump and be sure it's firmly seated, so it doesn't fall out.

15 Inspect the rubber seal around the sending unit opening. Make sure it isn't cracked or damaged and that it is correctly aligned.

16 Carefully guide the pump assembly through the hole; make sure not to get the sending unit's float caught on anything.

17 Seat the sending unit in the opening.

18 Install the lock ring, and tighten with the brass drift and hammer.

19 Place a floor jack under the tank assembly and raise it into the installed position. Don't attach the tank straps and bolt it up into position just yet. Find a good place to mount the fuel filter/regulator assembly; it needs to be in a safe spot that's protected from heat and any obstructions, such as the exhaust but as close to the tank as possible. The floorpan is the best place. It can be connected with self-tapping screws or, preferably, Nutserts or Riv-nuts.

20 Once in place, measure to see how long the flex lines back to the tank need to be, being careful not to put tension on them or make any kinks.

21 Once you have measured the line length, build an AN -6 line for each and install a standard male fitting on each end. Depending on the routing, the fitting may be straight, 45 degrees, or 90 degrees. Use straight fittings on all of them, but cut the lines so that there is plenty of length for a gradual curve.

22 Bolt the tank in place after you have attached the flex lines to the sending unit; it is easier with the tank out of the car.

23 Properly tighten the lines, and make sure they are well secured with Adel clamps or similar. Where they attach to the frame, drill and tap holes to use 10-24 button-head bolts.

Also, note which line is the feed; it is the larger one, on both the tank and the filter/regulator. The filter/regulator doesn't have AN lines. However, Russell and others make quick-disconnect AN -6 fittings that are offered in both 5/16-inch return and 3/8-inch feed. Therefore, you need one conventional 5/16-inch fitting (for the return) and two 3/8-inch fittings (one for the feed and one for the fuel rail on the engine).

You also need an additional 3/8-inch fitting for the front of the regulator, but this one is a quick-disconnect for the female end on the front of the regulator. This quick-disconnect end has a longer steel tube that pushes right into place; the design allows the filter/regulator to be easily disconnected and removed for replacement as needed.

24 Use a deadblow hammer to carefully tap the quick-disconnect fitting into place.

Quick Disconnect Fittings

If you're using the type with a plastic insert, be sure that the fitting is well seated on the flare, by tapping it into place with a deadblow hammer. If the plastic retainer isn't firmly gripping the flare, the fitting *will* come off. This is especially dangerous at the fuel rail, since high-pressure fuel will spray all over your headers. Eliminate the possibility entirely by using the newer screw-on type of fittings. ∎

The flex line is connected with a straight AN -6 fitting. (Photo Courtesy Matt McKahan)

 25 For the main pressure line, I recommend using steel, stainless steel, or aluminum hardline, rather than a flexible line. Some do not want to use aluminum fuel lines, but the 3/8-inch aluminum line is much easier to flare and bend than either mild or stainless steel. Aluminum lines work fine as long as they are properly secured to the frame so that they can't move around and work harden.

26 Start at the filter/regulator end of the line, and use a tube sleeve and tube nut to connect to the regulator.

27 Carefully route and bend the line to conform to the frame. It needs to be brought up on top of the frame near the firewall and be clear of the exhaust.

28 Flare that end, again with a tube sleeve and tube nut, and use a -6 union to connect another length of -6 hose with straight ends.

High-horsepower engines that require a high-volume fuel supply need some special equipment. The traditional method is to use a fuel cell and one or more large, in-line fuel pumps.

In this case, we need a conventional aftermarket EFI regulator. In most cases, high-capacity extruded aluminum replaces the stock fuel rails. The pressure line (or lines) feeds into the rails either directly or with a crossover. The return line is g-plumbed back to an EFI-specific regulator, which has high-pressure fuel requirements.

Typically, line sizes range from -6 to -10 (5/8-inch ID). This type of system works great for a drag car or occasional-use street car, but the pump is the weak link because it gets very hot and fails if run for long periods of time.

Electronic fuel controllers, such as the Aeromotive Billet Fuel Pump Speed Controller, help immensely with this issue. They reduce the voltage to the pump, but this doesn't completely solve the overheating problem with the fuel pump.

Aeromotive developed its Stealth fuel cells as a reliable solution. Essentially, the company took a high-quality aluminum fuel cell, and built in an A1000 or Eliminator fuel pump and a baffling system to keep the pump well covered. This system, with the Eliminator pump, is what will be running in the *GNXcess*.

Cooling System

You need a higher capacity cooling system and the requisite plumbing to properly cool an LS engine or high-performance small-block or big-block. The typical stock G-Body radiator is adequate for a stock engine. The stock radiator isn't suitable for the swap because the inlet and outlet locations on an LS engine are both on the passenger side, as opposed to being on opposite sides as with original radiators.

Because the layout and cooling capacity are not sufficient, an aftermarket aluminum radiator that is designed specifically for an LS engine swap is used. I have had excellent results with Griffin and Afco radiators.

You need a radiator that has correct inlet and outlet positions, but it should also have a provision for the steam line attachment. Some manufacturers don't include steam line fittings, so be sure you're getting a kit that has them. If necessary, you can tap and install a fitting in the steam line to the top of the water pump; I have done that and it worked fine. However, I recommend ordering the correct radiator because it is easier.

Beware of bargain-priced radiators found online because the quality may be suspect. Although some look like name-brand radiators, they often have much smaller coolant passages in the cores and do not cool a car effectively. Most also use epoxy rather than fully TIG welded cores, which makes repairs nearly impossible.

Many manufacturers offer their own shroud and dual-fan packages,

CHAPTER 11

The 525-hp LS3 was mocked up in our 1983 T-Type. The A/C box clears the headers with an LS3 installed, but in this case, we are deleting it and building a new firewall.

Oil pan clearance (even with the Musclecar pan) is tight to the crossmember, but nothing hits, even at full lock. The headers leave plenty of room for upper control arm removal and for alignments. Fabricating or modifying a pan to fit these cars is a thing of the past.

There is plenty of room for steering shaft clearance, even when using the stock rag joint. It does not contact the power steering box or any other vital components.

which is an effective cooling solution for a bolt-in swap. In most cases, you don't need a controller because the ECM (through its wiring) controls the fans. In any case, be sure the fans you use are effectively shrouded. The back side of the radiator should be almost entirely blocked off, so the fans force the hot air out.

Some manufacturers place louvers or small holes in the shroud to relieve some of the air pressure at speed, which doesn't impact cooling efficiency. But you must use a shroud with an electric fan so the engine does not overheat.

Swap Kits

Many would-be swappers do not attempt some swaps because of the obstacles to be overcome. Bolting in an engine and transmission is one thing, but many enthusiasts do not possess the welding and fabrication skills required for a custom swap. The easy solution is to use a swap kit.

I have used kit parts from BRP Hot Rods to complete the LS swap into many G-Body cars. BRP offers a line of engine mounts and matching transmission crossmembers, along with headers, oil pans, and many other needed items. The engineering is already done; just bolt them on, and you have a well-engineered swap with many of the hurdles already overcome.

Here are the general steps for performing your own G-Body LS swap with a BRP kit:

1. Remove the driveshaft, the original engine, and transmission.
2. Pull the powertrain, except for the rear end, from the car.
3. Remove the original engine's frame mounts. They are

BRP Swap Kit

BRP Hot Rods is a pioneer of LS engine swaps, offering dozens of kits. When developing a kit, the crew logically builds it around a fully optioned, original car to ensure that everything fits perfectly.

The company builds its own custom mounts that bolt directly to the original frame, using the factory holes. It also developed transmission crossmembers that work with virtually any GM or aftermarket transmissions, and that allows plenty of clearance for dual-exhaust systems. Any guesswork on driveline angle is eliminated, so you can leave the angle finder in your toolbox.

BRP always recommends a specific oil pan. In our case, it is the LH8 pan, which is the same as the Chevrolet Performance "Musclecar" pan kit. BRP did some additional machining on their version, for use with the newer displacement-on-demand engines, but this isn't necessary in most cases. Both kits include Hummer H3 pans, pan gasket, windage tray, pickup, dipstick and tube, oil filter (sometimes; General Motors isn't consistent on this), and all needed bolts. Individually from General Motors, all of this would cost well over $400, but the muscle car kit retails for around $180 at most Chevrolet Performance dealers.

First, BRP technicians mock up an engine with their mounts and the correct pan in the car, and then they send the car to Hedman Hedders' Husler racing division, which is in nearby Alpharetta, Georgia. Once there, the Hedman crew designs both mid-length and full-length headers in a variety of primary sizes for just about any application. As with most race headers, the collectors are removable; a single-bolt tab holds each together, and the pipes are simply slipped into place. Between this and the slip fit at the outlet of the collector, these headers are very easy to install.

BRP headers come with a very nice silver ceramic coating. Or they can be special ordered in bare steel if needed for a custom project. All come with high-quality flange gaskets, bolts, and oxygen sensor bungs, which are welded into the head pipes behind the collector.

---

basically a large chunk of rubber surrounded by steel and bolted to the frame with four bolts. Unfortunately, General Motors didn't see fit to use weld nuts, so you most likely have to use a combination of extensions and swivels to get to the nuts up inside the frame. This is much easier with the lower control arms removed, but it can be done with them in place.

4 Use the supplied Grade 8 hardware to loosely bolt the BRP mounts to the frame, and then fasten the engine side of the mounts to the engine. BRP indicates the correct side on the back of the mounts to avoid any confusion.

5 If the front end sheet metal has been removed, or you're dealing with a bare chassis (as in a typical full build), bolt the transmission to the engine and install it as a unit.

Otherwise, install the engine first, and then install the transmission later from underneath. This saves you a lot of frustration, and even though initially it is more work, it is faster and easier in the long run. It also minimizes the chance of damaging the firewall or A/C box.

6 Use a lift chain or a conventional engine tilter that uses separate chains to lift the replacement engine for installation; remember that the LS engine has a factory plastic intake that's easily marred, so remove it beforehand. (If you have a carb-equipped LS engine, you can attach an engine hoist to a lift plate on the carburetor flange.)

---

1 Lower the engine over the frame mounts. You need a little help to do this. Have someone underneath, and two people above to control the hoist and move the engine as needed.

2 Slide in the motor mount through bolts, one at a time, and loosely thread on the nuts. It may take a little jockeying of the engine to do this.

3 Once the transmission is bolted to the engine, bolt the supplied urethane transmission mount to the transmission.

4 Position the new crossmember under the frame rails and bolt it to the transmission mount.

5 Jack up the transmission until it reaches the proper position in the tunnel; that is when the bolt

CHAPTER 11

TPIS "Happy Hooker" Engine Lift Plate

For a typical LT1 engine swap, the chassis is raised on a lift and the engine is dropped out the bottom of the car. The entire front suspension and engine are removed at the same time.

Doing it this way took too much time and effort between rounds on the World Challenge race circuit, so TPiS developed the "Happy Hooker" to bolt to the LT1's intake. It is a very simple design that features a long, flat plate with a swivel in the middle, and a curved rod with a steel loop on the end; it works extremely well.

The plate comes with several patterns drilled in it, and several more can be drilled to use it on any other engine removal or installation.

You remove the valley cover, and install the Hooker using the original bolts. You could use a chain if you prefer, but be sure to take the intake off first and bolt the chain to the front of the cylinder head on one side, and to the rear of the opposite head.

---

flanges on each side of the crossmember are resting evenly on the frame rail.

6 Mark the holes with a punch, or use the holes in the crossmember as a guide to drill holes in the bottom of the frame rail.

7 Once all four holes are drilled, thread in the bolts to ensure that they fit.

8 Unbolt the crossmember and insert the ends above the lower flange of the frame rail and bolt it in place.

9 Reinstall the transmission mount bolts and carefully tighten them. You can also tighten all the engine mount bolts at this time.

BRP's bolt-together headers are very easy to install, even the large-tube types. When installing typical headers, it is easier to leave the mounts loose, but that's not necessary for BRP headers. Even the large 1⅞-inch primaries have no clearance issues, and their installation is very simple.

Here are the general steps for installing BRP headers:

1 Disassemble the headers and loosely bolt the flange to the cylinder head with a bolt on each end, sandwiching the supplied gasket.

2 Loosely install the other bolts.

3 Use a bit of high-temperature copper RTV inside the front of the collector, and slide the collector over the primary tubes.

4 Reinstall the bolt with a dab of thread locker.

5 Go back and torque the header bolts to spec (18 ft-lbs), and this part is complete.

Gauges

Regardless of the method of fueling the engine, you're going to need gauges to monitor it. A common question is, "How will the gauges work with this new engine?"

The speedometer is the biggest mystery for most, but it's fairly simple to upgrade if you know what type of speedometer you have. The original and some aftermarket speedometers use a cable to drive them mechanically. A mechanical speedometer uses the original speedometer cable and is compatible with non-electronic transmissions.

If you are using an electronic transmission, determine what type of cable the speedometer uses. Most aftermarket versions use the old-style, screw-on cable end, but most originals in G-Bodies use a clip-on end. With this information, you can order a speedometer conversion box, such as a Cable-X from Abbott Enterprises. The box converts the VSS signal from an electronic transmission to a mechanical cable drive.

It may be cheaper to replace a mechanical speedometer with an electronic one if it is an aftermarket mechanical type and the manufacturer offers an electronic version. If the speedometer and the transmission are both electronic, everything will work as-is.

If you are using an original tachometer or an older aftermarket one that isn't adjustable for different engine sizes, you need a tach adapter, such as one from Auto Meter. For an aftermarket tach that can be configured for different engines (either by clipping a wire loop or resetting DIP switches on the back), set it for a four-cylinder engine and wire it according to the manufacturer's instructions.

Mechanical, electrical, original, aftermarket oil pressure, and water temperature gauges typically don't adapt easily to LS engines with what comes in the package. Most gauges are designed for NPT, and everything on these engines is metric. Some people drill and tap new holes, and use NPT fittings. However, unless you already have the NPT taps on hand, it's usually less expensive and easier to get the correct gauge adapters rather than source these hard-to-find and expensive fittings. I buy them as an assortment from Equus, and usually keep several packs on hand. Auto Meter makes adapters in several metric to NPT sizes also, but they are more expensive because they are sold individually.

Accessories

The typical street engine is equipped with an alternator, power steering pump, and more often than not, an air conditioning compressor. Installing all these in your car, while clearing the front hood and crossmember, can be an issue. The following is some helpful information.

Brackets

Crate engines don't come with accessories or accessory brackets, and they usually are set up with the Corvette front hub. Takeout (or salvage) engines often do have the complete front engine accessory drive, but in many cases it isn't usable.

For example, the Trailblazer SS is a popular engine source because it comes with an LS2 that's calibrated for trucks. However, the taller truck intake manifold and accessory brackets are a little tall for use in a G-Body. Many suppliers provide these engines with the original accessories, minus the A/C compressor (which isn't compatible with most swaps anyway). The truck-style accessory brackets space everything out farther than the Corvette brackets do, and use a correspondingly longer front hub. In the past, swappers had few engine accessory options. Often, they replaced the entire stock front dress, hub, and water pump on a truck engine with F-Body or Corvette parts. This works, but it's expensive and a pain to track everything down. Chevrolet Performance offers a complete FEAD kit to make the accessory conversion much easier. The kit is much less expensive than buying each component individually from General Motors. It uses the stock Corvette A/C compressor, which sits low on the passenger side and usually requires frame notching for clearance.

In my opinion, installing an accessory bracket set from Kwik Performance is the best solution. Kwik's bracket kits replicate the basic Corvette accessory layout, but the A/C bracket is moved to the upper right-hand side, and a more swap-friendly Sanden compressor is used. Kwik kits include needed spacers so the accessories are properly spaced and you don't have to change hubs.

As a result, the stock water pump and hub can be reused, which saves money. The brackets accept any of the available LS engine alternators

This LS2 installed in a 1969 Camaro shows several items common to swapping an LS engine into a G-Body. There are Kwik Performance accessory brackets, BRP/Musclerods mounts and headers, and KRC power steering reservoirs. The air-intake plumbing was assembled from Spectre Performance pieces, and the radiator was designed specifically for an LS engine swap. The inlet and outlet were positioned on the passenger side, and there was a fitting for the steam hose. Bulldawg Musclecars built the fan shroud package.

CHAPTER 11

Accessory brackets used to be more of an issue, especially on A/C-equipped swaps, but Kwik Performance solved the problem with its affordable bracket kits. The existing hub and water pump can be used regardless of engine choice, with one exception: Truck engines with passenger car manifolds and cable-driven throttle bodies need to switch to a late-Camaro water pump.

Holley now has its own line of cast-aluminum brackets, along with just about everything else needed for a swap. Holley's A/C bracket uses a serpentine-style R-4 compressor. This was installed on very few stock G-Bodies, but later retrofitted to many as part of a serpentine system swap. Holley also has a version to work with a more conventional Sanden-style compressor.

and power steering pumps, though truck-style power steering pumps with internal reservoirs need to be converted to use an external reservoir. The style of alternator plug on the harness determines which alternator is compatible. Most early applications use a four-pin connector while most later ones use a three-pin connector.

Power Steering

Plumbing the power steering system is fairly easy, but the steering box itself must be compatible. OEM steering boxes on early G-Body cars have conventional flares, rather than the later (built after 1979) O-ring style.

The return hose doesn't require a certain fitting style because you re-use the metal part of the return line. Simply measure the needed length of power steering hose and buy a piece of bulk hose from an auto parts store to connect it to the return port on the pump. The easiest option is to use an off-the-shelf power steering hose for the pressure side.

Hoses from late-model 4.3-liter S10 trucks have been installed onto G-Body project cars, and many swaps just need a little re-bending of the hardline. Typically, a local hydraulic shop can make a hose according to your specs. They have the ability to build a true high-pressure, stainless, braided hose that is far stronger than anything you can buy and build at home, and it's often cheaper than a parts store hose that "almost" fits.

You can use a KRC power steering reservoir that is TIG-welded aluminum. They come with a -10 fitting for the feed line to the pump and a -6 return. I suggest you use a standard -10 90-degree fitting and correspondingly sized hydraulic hose, plus a barb fitting with a clamp (I prefer Gates PowerGrip clamps) at the pump end.

Air Conditioning

Your particular A/C setup varies according to engine package and accessory layout. For example, the Sanden compressor requires custom hoses. Some G-Body models, including many Olds G-Body cars, have a condenser with a passenger-side hose connection that's much cleaner than the more common driver-side connection.

Carefully measure and route the hoses, so you get it right and do not have to do it a second time. Often the length and route are simply personal preference. Once you have a rough measurement, take the original hoses and the Sanden compressor to an A/C shop and have them make up lines to your specs.

CHAPTER 12

DRIVING IN THE LAP OF LUXURY
PERFORMANCE INTERIOR UPGRADES

A well-appointed and well-laid-out interior greatly improves comfort, but it also improves control and the overall driving experience. In stock form, most G-Bodies, even the performance-oriented models, leave a lot to be desired in interior equipment and appointments. Many had, by today's standards, undesirable color combinations and unappealing materials that give the car a cheap look. Lots of plastic, cheap vinyl and cloth, and fake woodgrain all fall into this category. Instrumentation in most models was virtually non-existent, relying instead on warning lights. This is especially disconcerting when compared to the average commuter car today, which typically has real gauges that include a tachometer. Seating in the average G-Body is more akin to the furniture in your grandmother's living room than anything in a modern performance vehicle.

Gauges

Since this book is about performance and not restoration, let's start with one of the areas most sorely lacking in the G-Body: the gauges. The average G-Body had a rather stodgy horizontal-style mechanical speedometer, and it only went up to 85 mph. That seemed like enough at a time of the 55-mph speed limit, but 150 mph or more can be found in the average performance car of today. Even the turbo Buick models had this setup, and I can speak from experience that it just wasn't enough. Buick remedied that on the GNX, with an ASC/McLaren gauge cluster full of Stewart-Warner gauges, but if you aren't one of the fortunate 547, you have your work cut out for you.

Most people simply remove the existing gauges, and install a piece of aluminum or stainless steel in its place, with appropriate-size holes cut

This Malibu has an exceptional performance interior. It features a roll bar with swing-out door bars, leather seats, and console from a later-model F-Body; Auto Meter Phantom Series gauges in custom dash inserts; Simpson Cam-Lock five-way harnesses; aftermarket shifter; and custom steering wheel. These are additions that can be made gradually, as time and money allow.

CHAPTER 12

When starting a G-Body build, an average interior often looks like this, stock and in very good shape. It already has bucket seats, a console, and the coveted floor-shift steering column. The fabric surfaces can easily be recovered in a more attractive color, the carpet replaced, and all the various plastic and vinyl interior trim and dash dyed.

The owner of this vehicle has added stainless-steel inserts to the dash for a little extra flash, and to block off the old vents. The entire HVAC system has been removed from this Cutlass, so they were no longer needed.

There are many gauge choices, even within the same brand. These are from Auto Meter.

The GNXray's interior uses mostly stock components, but the seats have been recovered in suede and leather, and the interior plastic and vinyl have been dyed black. The gauge cluster features Auto Meter Phantom Series gauges.

for your gauges of choice. There are also pre-made panels available from a variety of sources, if you don't want to fabricate a panel. Find a brand and style of gauge you like, with a couple things in mind.

One, be sure that the gauge manufacturer offers a gauge for every function that you want. For example, some may not offer a boost gauge, or an air/fuel ratio gauge. Rather than use mismatched gauges, find a line of gauges that has everything you need.

Two, consider the method of operation for the gauges: electric or mechanical. Electric gauges got a bad rap early on for inaccuracy and a limited sweep compared to a mechanical gauge, but modern electric gauges have come a long way. Most manufacturers tell you that their electric gauges are just as accurate as the

The interior of this immaculate Monte Carlo SS drag car features a mostly stock dash, but has auxiliary gauges and a switch panel replacing the stock HVAC controls and the radio. A lightweight drag-style column with a quick-release steering wheel is the only major modification. The stock instruments remain, but are augmented by a shift light (rather than the typical monster-style tachometer).

136 GM G-BODY PERFORMANCE UPGRADES 1978–1987

DRIVING IN THE LAP OF LUXURY

Auto Meter gauges are very popular in G-Bodies. These are Ultra-lights. The shifter is a Hurst Quarter Stick.

The factory gauges have been augmented with this panel from Auto Meter. The transmission temperature gauge is critical in a car that is driven hard, especially on autocross and road courses. (Photo Courtesy Michael Weddle Photography)

mechanical ones, and in many cases the sweep range is increased. In any case, they are easy to install, and there isn't a chance of liquid (oil or fuel) leaking into the interior due to a bad line or connection.

Most mechanical gauges are available with an isolator, which keeps flammable liquids out of the cockpit, but they still add complexity and cost to the installation. For this reason, I prefer to stick with electric gauges whenever possible.

Speedometer

I recommend going with a speedometer that works best for your existing transmission. Depending on what transmission you are using, and what year vehicle it is from, the speedometer may be driven by a mechanical cable or by a signal from the VSS. If yours is set up for a mechanical speedometer, use a mechanical one. They are usually much cheaper, and work fine. If you are using a later transmission (say in conjunction with an LS-series swap), a mechanical speedometer won't work.

There are three choices: use a conversion box with your existing mechanical speedometer; switch to an electronic speedometer; or go to a GPS-driven speedometer.

I use a conversion box only in cases where I want to retain the original speedometer, since an aftermarket electronic version is generally less expensive. GPS speedometers are gaining popularity, though, since they are very easy to hook up (power and ground), and are very accurate.

Tachometer

The main consideration when choosing a tachometer is if you are running an engine equipped with coil packs (any LS-series engine or a conversion on another engine that uses the same coils). LS-series coils use a pulse signal from four cylinders, rather than eight, so a tach that is calibrated for a V-8 won't work.

There are three solutions: purchase an aftermarket tach that can be set for a four-cylinder application, use a tach adapter such as the Auto Meter 9117, or have a tuner change the output signal to match your tachometer's requirements.

Pressure and Temperature

Hooking up the pressure and temperature gauges is pretty straightforward, but sometimes the adapters that come with the gauges aren't the ones needed for your engine, or your factory-style sending units are the wrong size to go with your later engine (common in LT1 and LS-series swaps). Gauge adapters are available from Auto Meter and others, and in kits from Equus. These often solve the problem. In some cases, though, due

These gauges are from Dakota Digital. They are installed in Dan Howe's 1984 Monte Carlo SS.

Auto Meter offers numerous styles of gauges, with essentially the same layout. They are very popular for racers and street car enthusiasts alike.

GM G-BODY PERFORMANCE UPGRADES 1978–1987

to your own individual parts choices, it may be necessary to get creative.

Most engines have coolant plugs in various locations: in the cylinder heads, in the intake manifold, even in the block itself. On most GM engines the factory oil pressure gauge typically connects behind the intake manifold, near the bellhousing flange. In most cases you can also tap the oil filter housing. For LS-series engines, a small removable cover above the oil filter can be drilled and tapped if the standard location behind the intake doesn't work for your swap.

Wiring Harness

Because these vehicles were new when I was in high school, I sometimes forget that the newest ones are 25 years old, and many of the things that weren't considerations in modifying them back then are now major issues. A perfect example is the wiring harness. Even in an unmolested car, the wiring is often brittle and cracked. The typical car has had multiple stereo components, alarms, and auxiliary lighting installed, and various ham-fisted repairs made over the years. It's a good idea to replace all the wiring (be sure to save all the factory connectors from the old harness; they make connections to the lights and other accessories much easier).

Replacement harnesses have long since been discontinued by General Motors, and the aftermarket doesn't have any reproduction harnesses available either. This is an issue for the restorer of these cars, but for someone who is modifying them, the best solution is to use an aftermarket, universal wiring harness such as those sold by Painless Wiring and American Auto Wire. This way, each circuit can be tailored to your specific needs, rather than having several add-ons piggybacked to a single fuse. Most change out the factory gauges for aftermarket ones, add electronics such as switch panels and stereos, and in many cases change out the engine to another type. Starting with a clean sheet of paper on the wiring just makes sense.

Installing one of these kits is fairly simple, but doing it correctly is *very* time consuming; 30 to 40 hours isn't unusual. The process usually begins by mounting the fuse block in an easily accessible location (typically under the dash), and separating bundles of wires according to what part of the car they are going to. They are secured temporarily with twist ties.

Once the routing is determined, each strand is run through appropriately sized shrink tubing (available in bulk from places such as Fry Electronics or Del City), and solder and shrink tubing are used on all the connections. Each branch of the harness is secured with small clamps, to avoid any stress or abrasion, and wire is tied where necessary.

This is a great opportunity to clean up the overall appearance of your car, by routing wires more neatly and out of sight where possible. You can increase reliability at the same time.

Steering Wheel

G-Body performance models, such as the Grand National and Monte Carlo SS, came with leather-wrapped steering wheels that were both stylish and comfortable. Even if your car came with one, though, chances are good that it isn't in the best shape after more than 25 years of usage. Most other G-Bodies were equipped with small-rimmed, plastic wheels that are often cracked and discolored, and frankly have no place in a performance vehicle.

Dan Howe replaced his Monte's steering wheel with this leather-wrapped unit from Grant, and it features a Chevy Bow Tie in the center. Of all the steering wheel manufacturers, Grant offers dozens of styles and the adapter kits to install them, which has made these steering wheels very popular. They are available anywhere from your corner parts store to large mail-order suppliers, such as Summit Racing.

This is a small sampling of Grant steering wheel styles. There is a style and size to fit virtually every taste and budget, and the adapter kits are available to install them in anything.

DRIVING IN THE LAP OF LUXURY

Flaming River is another source for steering wheels. The company offers several styles; these are mostly billet aluminum.

Fortunately, this is an easy and relatively inexpensive swap, requiring only a steering wheel puller (available for loan at most auto parts stores, if you don't have one) and a few minutes of your time.

When buying a wheel, be sure to purchase the appropriate adapter kit for your car, since the wheels themselves are universal.

Shifter

If you are running an automatic transmission, and don't like the column shift found in most G-Bodies, a floor shifter is an easy upgrade. Virtually any factory cable-operated automatic floor shifter can be installed; it doesn't have to be a G-Body shifter. F-Bodies from 1982–2002 are a good source, and the consoles can be added

Billet Specialties offers several styles of steering wheel, including this Chicayne wheel. They are made of billet aluminum and are available with either a full or half leather wrap.

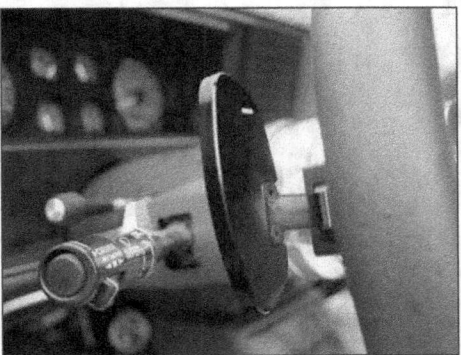

Howe recently added a TCI paddle shifter so he can shift the transmission without taking his hands off the steering wheel. It provides improved control on the autocross course. This type of unit requires an electronic transmission with an add-on controller to power it.

The Hurst V-Matic shifter (shown) is very similar to the B&M Megashifter. Both are available for nearly any 3- or 4-speed automatic transmission.

easily as well, by welding in the mounting brackets from a donor car.

There are dozens of aftermarket shifters available, so the chances of finding something that suits your

It doesn't look like it, but Doug Lutes' Sic Monte interior still uses a lot of stock components. Even the shifter is original and does its job well. The Auto Meter gauges below the ashtray and the blockoff plate for the HVAC controls are obvious changes. Everything else is stock Chevrolet.

Dan Howe's Monte Carlo SS came with this B&M Megashifter. It is a little redundant now because he is running a TCI paddle shifter, but it's always nice to have a backup plan in case of an electrical issue.

CHAPTER 12

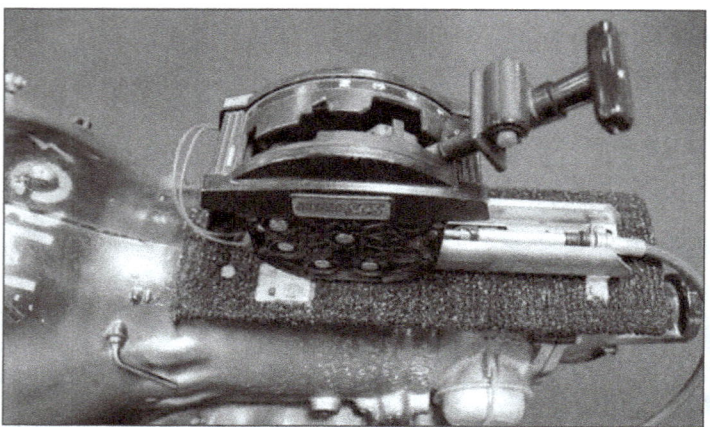

The Turbo-Action Cheetah SCS shifter isn't as well known as other shifters on the market, but I prefer it. It is very easy to use and very durable.

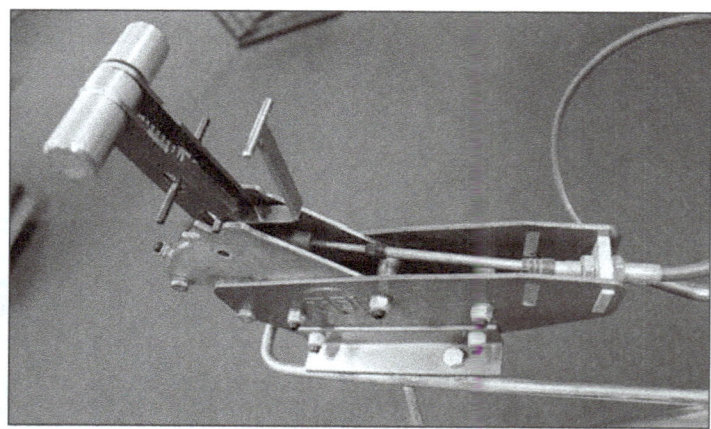

This TCI Lightning is a compact shifter and a good choice for a G-Body drag car. It is set up for the 2-speed Power-Glide.

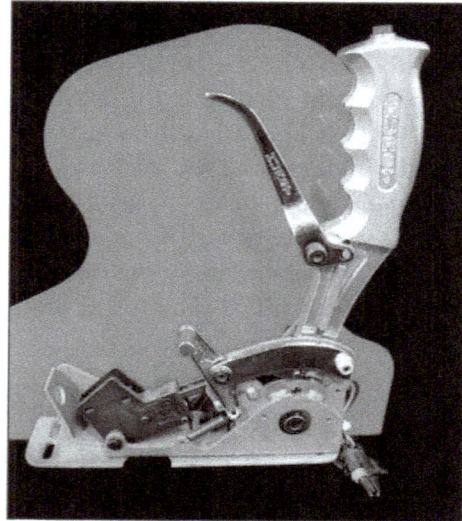

The Hurst Quarter Stick has been a popular shifter for many years, especially in drag-oriented cars. They are available with either a conventional round knob or a pistol grip version (shown).

purposes and style are good. They also have the advantage of a positive lockout for reverse and park.

Seats

G-Bodies came with a variety of seating. The 60/40 split-bench seats are the most common, with conventional high-back bucket seats coming in a close second. Some cars came with a wider, low-back bucket seat (mostly Regal T-Types), but these are fairly rare.

High-back bucket seats are the most desirable for a performance application. They are lightweight and fairly supportive. Reproduction seat covers are available, or you can have custom covers made. I lean more toward the latter, since an upholsterer then has the option of adding foam to the bolsters. This is a great way to gain support, especially if you like to push your car in the curves. This also helps to retain more of a stock look, if that is important to you.

If your car didn't come with buckets, and you want them, find a donor car to provide the mounting brackets. The seat tracks are specific to the G-Body, so you definitely need them. General Motors made most of its seats with the same bolt pattern at the track attachment points, so you aren't limited to G-Body seats.

Virtually every car from the G-Body era to the present had a bucket-seat option, and they bolt to G-Body tracks. I have even used bucket seats from a four-door Pontiac 6000 wagon. It had a perfect set of PMD buckets (also known as Viscount, or Knight Rider seats), but since it was a four-door front seat the backs were fixed. I got around this by robbing the seat-folding mechanism from another set of seats (in an F-Body, but virtually any works), which was a direct bolt-on. This is a good trick to know, since bucket seats in four-door models are often in much better shape.

Want aftermarket bucket seats or a racing bucket? Most come with tracks or have tracks as an option, but many of them have only a "universal" track. In cases such as this, it may be necessary to either alter the supplied tracks or adapt the original seat tracks to the new seats. Depending on the seats and tracks in question, one of these options may be easier than the other. The important thing is to make sure that they are properly secured to the floor.

Whenever possible, use all of the factory mounting points for the original seats. They are properly reinforced, and hold the load adequately. *Never* secure the seats with bolts and large washers. In a collision, these washers rip right through the sheet-metal floorpan, and chances for

DRIVING IN THE LAP OF LUXURY

These Kirkey aluminum seats are ideal for serious racers. The supplied brackets are easily adapted to mount to the stock seat mounting points on a G-Body. The new G-Force harnesses are ready to install. (Photo Courtesy Doug Lutes)

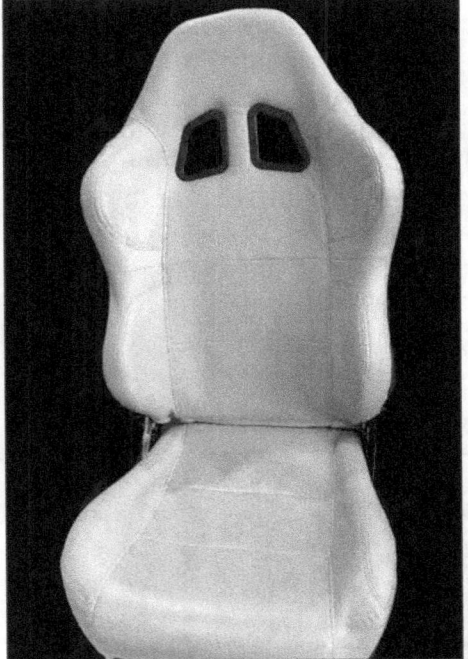

These Summit Sport Seats are suitable for nearly any project and are very reasonably priced. Universal brackets are available separately.

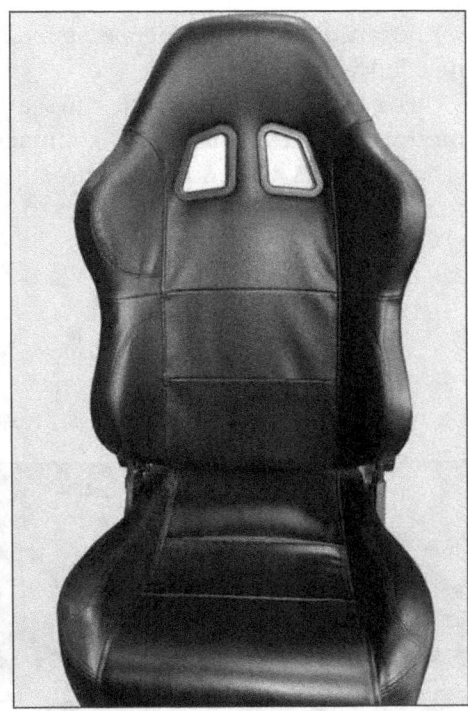

Summit Sport Seats are also available in black.

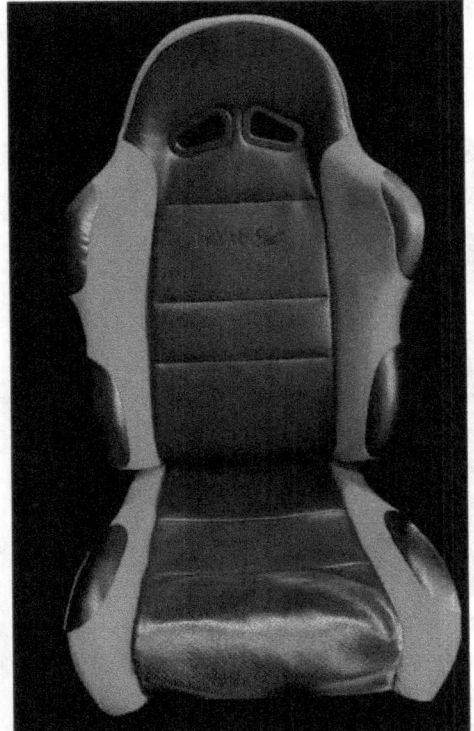

This Scat ProCar seat is of similar size and styling to the Summit Sport Seats but with more color choices, including this two-tone example.

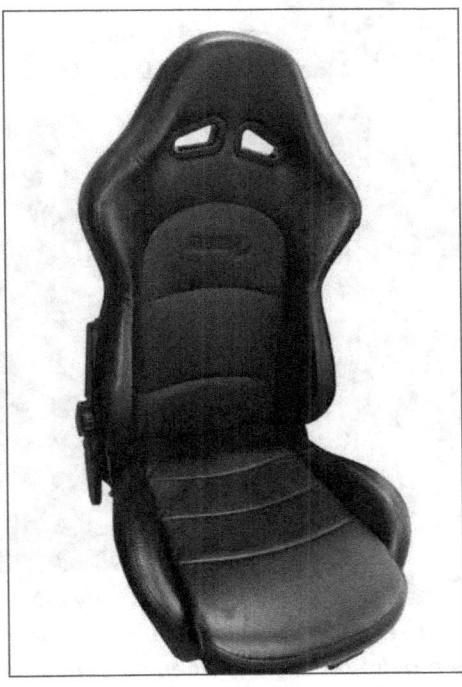

Most enthusiasts choose traditional black upholstery, as found on this Scat ProCar seat. This is one of the most commonly used aftermarket seats in current production.

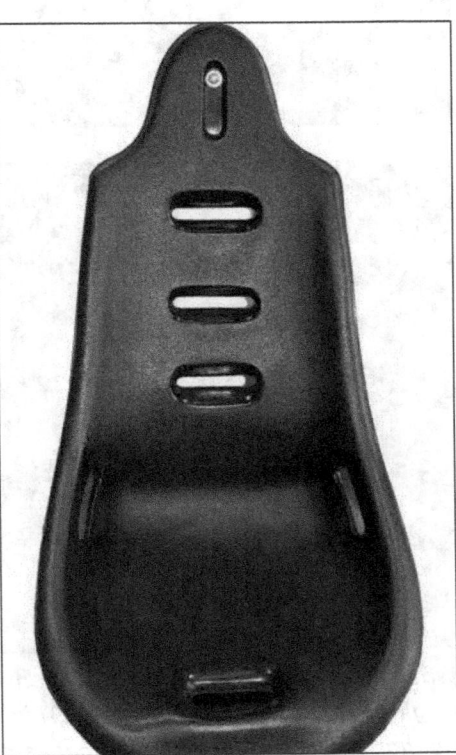

Summit's Poly Pro seats are a good budget choice for a drag car at less than $40 each. This doesn't include mounting brackets.

CHAPTER 12

serious injury or death increase exponentially.

For a race application, tie the seat mounting directly into the tubular crossmembers that connect directly to the frame rails. (This is done in the *GNXcess*.)

Safety Restraints

The subject of seat belts needs to be carefully considered. Most states require seat belts for any vehicle originally equipped with them, and even if your state doesn't, you should have them. If yours are damaged, get a set from a donor car or have them re-webbed and restored by a company such as Ssnake-Oyl Products. This is also a great way to go if the seat belts are the wrong color for the new interior.

Any vehicle with a roll bar must have racing harnesses to supplement

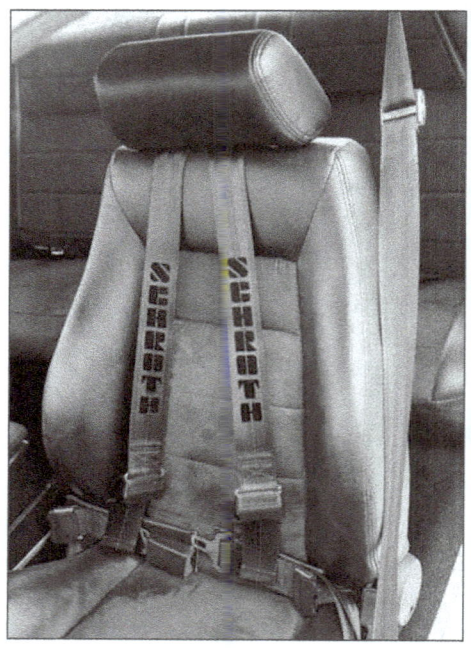

The seats in Scott Walkowiak's GNXray are the stock T-Type buckets, but they have been recovered with suede inserts and leather bolsters. In addition, they are equipped with Schroth harnesses. Note that the stock shoulder harness has been retained for street use and legal compliance.

G-Force harnesses are securely mounted to the roll bar. This shows the ideal mounting points. (Photo Courtesy Doug Lutes)

When adding aftermarket seats of any kind, use the original seat mounting locations. This may require modifications to the tracks in some cases but is necessary for safety. These buckets are from Kirkey, and are designed to keep the driver firmly in place under racing conditions. They are very supportive, but not the most comfortable for long trips. (Photo Courtesy Michael Weddle Photography)

High-quality aftermarket bucket seats modernize any interior, and they provide much-needed support for the driver when pushing the car to its limits. A set of five-point harnesses, such as these from G-Force, is a wise and often required addition, if your car has a roll bar or cage. Without a bar, I recommend retaining the factory belts because mounting these safely without one is nearly impossible. (Photo Courtesy Doug Lutes)

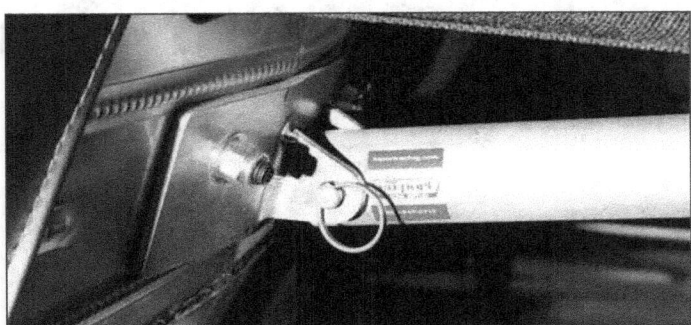
In a competition car, you want to have a support running from the cross bar of the roll cage to the seat back so that it has rigidity in a crash. Some only have a plate that rests against the seatback, but the supports on these Kirkey seats are bolted in place for extra safety. They can be easily unbolted for service. (Photo Courtesy Michael Weddle Photography)

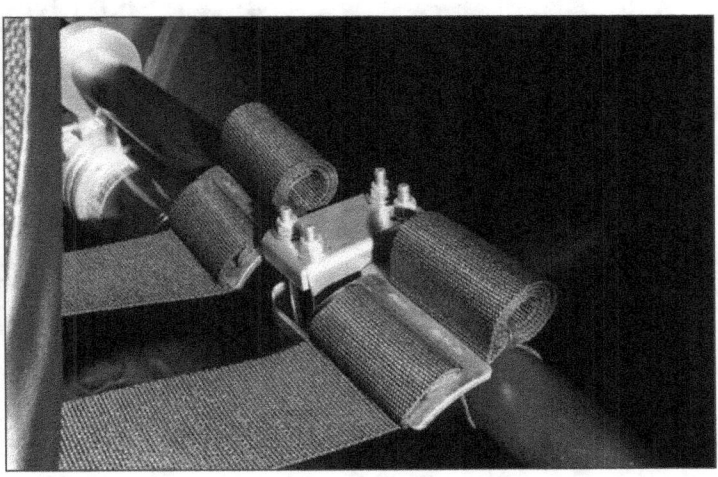
The G-Force harnesses are securely mounted to the roll bar. This is Doug Lutes' new setup with his Kirkey seats. (Photo Courtesy Michael Weddle Photography)

the original belts, and I use them any time the vehicle is driven even if technically the harnesses aren't DOT legal. A factory-style seat belt isn't designed to keep your head away from a roll bar, and a serious head injury or death could occur. Personally, I'd rather risk getting a ticket for improper belts than not maintaining safety for myself and my passengers.

The preferred method for securing five-point harnesses is to attach them to the roll bar, behind the seat, and to the chassis itself. The mounting points need to be able to withstand a 3,500-pound shock load (according to Simpson Performance Products, a leader in safety equipment). I trust nothing less than sandwiched steel plates, above and below the floorpan, with a welded nut in the center and the entire assembly welded in. If possible, tie them directly into a chassis member (as I described for the seats). Follow the instructions provided with your harnesses explicitly, as incorrect mounting can lead to spinal injuries in a crash. I highly recommend that any vehicle being used for racing have some type of head and neck restraints, such as the Hybrid Pro system from Simpson, or a comparable Hans device. Most require a compatible helmet, which adds to the cost, but are well worth it in my opinion.

Any car, especially a performance car, should have a fire extinguisher in the passenger compartment. Be sure that you securely mount it because in a crash this becomes a heavy projectile that could cause injury or death. It should also be easily accessible to the driver. (Photo Courtesy Michael Weddle Photography)

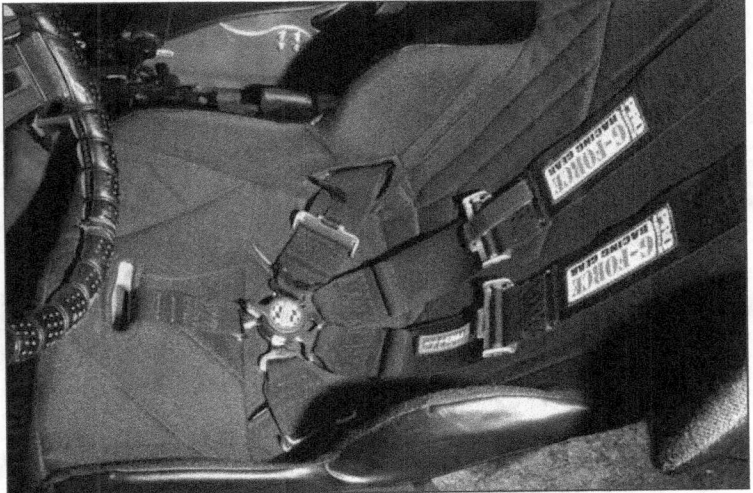
These G-Force Pro Series five-point harnesses use a cam-lock retainer, which is easier to use than the old-style buckles. This is well worth the additional cost over standard harnesses. (Photo Courtesy Michael Weddle Photography)

Source Guide

Abott Enterprises
901 W. 4th Ave.
Pine Bluff, AR. 71601
800-643-5973
www.Abbott-tach.com

Baer
2222 W. Peoria Ave.
Phoenix, AZ 85029
602-233-1411
www.baer.com

Barnett Performance
465 Memorial Dr. SE
Atlanta, GA 30312
800-533-1320
www.barnettperformance.com

BMR Suspension, Inc.
928 Sligh Ave.
Seffner, FL 33584
813-986-9302
www.bmrsuspension.com

BRP Hot Rods
5849 Rogers Rd.
Cumming, GA 30040
770-751-0687
www.brphotrods.com

Bulldawg Musclecars
4659 S. Main St.
Acworth, GA 30101
770-975-8980
www.bulldawgmusclecars.com

Chris Alston's Chassisworks
8661 Younger Creek Dr.
Sacramento, CA 95828
916-388-0288
www.cachassisworks.com

Day's Chevrolet
545 Hwy 15 S.
Jasper, GA 30142
877-641-0949
www.dayschevroletjasper.com

Detroit Speed & Engineering
185 McKenzie Rd.
Mooresville, NC 28115
704-662-3272
www.detroitspeed.com

Energy Suspension
1131 Via Callejon
San Clemente, CA 92673
949-361-3935
www.energysuspension.com

Global West Suspension
655 S. Lincoln Ave.
San Bernardino, CA 92408
909-890-0759
www.globalwest.net

Holley Performance Products
1801 Russellville Rd.
Bowling Green, KY 42101
270-781-9741
www.holley.com

Hotchkis Performance
8633 Sorensen Ave.
Santa Fe Springs, CA 90670
888-735-6425
www.hotchkis.net

KORE3 Industries LLC
32206 Tangent Dr.
Tangent, OR 97389
541-924-5673
www.kore3.com

Kwik Performance, Inc.
417-955-1467
www.kwikperformance.com

Metco Motorsports
109 North Park Drive
Anderson, SC 29625
864-332-5929
www.metcomotorsports.com

Modern Vintage Systems
317-224-6689
www.modernvintagesystems.blogspot.com

Art Morrison
5216 7th Street East
Fife, Washington 98424
888-640-1506
www.artmorrison.com

Mike Norris Motorsports
9402 Uptown Drive, Suite 1500
Indianapolis, IN 46256
407-616-2518
www.mikenorrismotorsports.com

RideTech
350 S. St. Charles St
Jasper, Indiana, 47546
812-481-4787
www.ridetech.com

Savitske Classic & Custom Hot Rod Shop
2059 Route 212
Coopersburg, PA 18036
610-381-6100
www.scandc.com

Schwartz Performance
1115 Rail Drive
Woodstock, IL 60098
815-206-2230
www.schwartzperformance.com

Spohn Performance
494 E. Lincoln Ave
Myerstown, PA 17067
888-365-6064
www.spohn.net

Street Rod Designs
4659 S Main Street
Acworth, GA 30101
801-471-3226
www.streetroddesigns.com

Summit Racing
1200 Southeast Ave.
Tallmadge, OH 44278
800-230-3030
www.summitracing.com

Tech AFX, Inc.
P.O. Box 252414
West Bloomfield, MI 48325
877-355-0137
www.techafx.com

TPIS
4255 Creek Rd
Chaska, MN 55318
952-448-7230
www.tpis.com

Wilwood
4700 Calle Bolero
Camarillo, CA 93012
805-388-1188
www.wilwood.com

www.ingramcontent.com/pod-product-compliance
Lightning Source LLC
Chambersburg PA
CBHW081451070526
44586CB00019B/2301